NUREG-1807

Probabilistic Fracture Mechanics – Models, Parameters, and Uncertainty Treatment Used in FAVOR Version 04.1

I0488590

Manuscript Completed: May 2006
Date Published: June 2007

Prepared by
M. EricksonKirk, B.R. Bass, T. Dickson,
C. Pugh, T. Santos, P. Williams

Division of Fuel, Engineering and Radiological Research
Office of Nuclear Regulatory Research
U.S. Nuclear Regulatory Commission
Washington, DC 20555-0001

ABSTRACT

During plant operation, the walls of reactor pressure vessels (RPVs) are exposed to neutron radiation, resulting in localized embrittlement of the vessel steel and weld materials in the core area. If an embrittled RPV had an existing flaw of critical size and certain severe system transients were to occur, the flaw could very rapidly propagate through the vessel, resulting in a through-wall crack and challenging the integrity of the RPV. The severe transients of concern, known as pressurized thermal shock (PTS), are characterized by a rapid cooling (i.e., thermal shock) of the internal RPV surface in combination with repressurization of the RPV. Advancements in our understanding and knowledge of materials behavior, our ability to realistically model plant systems and operational characteristics, and our ability to better evaluate PTS transients to estimate loads on vessel walls led the U.S. Nuclear Regulatory Commission (NRC) to realize that the earlier analysis, conducted in the course of developing the PTS Rule in the 1980s, contained significant conservatisms.

This report, which describes the technical basis for the probabilistic fracture mechanics model, is one of a series of 21 other documents detailing the results of the NRC study

FOREWORD

The reactor pressure vessel is exposed to neutron radiation during normal operation. Over time, the vessel steel becomes progressively more brittle in the region adjacent to the core. If a vessel had a preexisting flaw of critical size and certain severe system transients occurred, this flaw could propagate rapidly through the vessel, resulting in a through-wall crack. The severe transients of concern, known as pressurized thermal shock (PTS), are characterized by rapid cooling (i.e., thermal shock) of the internal reactor pressure vessel surface that may be combined with repressurization. The simultaneous occurrence of critical-size flaws, embrittled vessel, and a severe PTS transient is a very low probability event. The current study shows that U.S. pressurized-water reactors do not approach the levels of embrittlement to make them susceptible to PTS failure, even during extended operation well beyond the original 40-year design life.

Advancements in our understanding and knowledge of materials behavior, our ability to realistically model plant systems and operational characteristics, and our ability to better evaluate PTS transients to estimate loads on vessel walls have shown that earlier analyses, performed some 20 years ago as part of the development of the PTS rule, were overly conservative, based on the tools available at the time. Consistent with the NRC's Strategic Plan to use best-estimate analyses combined with uncertainty assessments to resolve safety-related issues, the NRC's Office of Nuclear Regulatory Research undertook a project in 1999 to develop a technical basis to support a risk-informed revision of the existing PTS Rule, set forth in Title 10, Section 50.61, of the Code of Federal Regulations (10 CFR 50.61).

Two central features of the current research approach were a focus on the use of realistic input values and models and an explicit treatment of uncertainties (using currently available uncertainty analysis tools and techniques). This approach improved significantly upon that employed in the past to establish the existing 10 CFR 50.61 embrittlement limits. The previous approach included unquantified conservatisms in many aspects of the analysis, and uncertainties were treated implicitly by incorporating them into the models.

This report is one of a series of 21 reports that provide the technical basis that the staff will consider in a potential revision of 10 CFR 50.61. The risk from PTS was determined from the integrated results of the Fifth Version of the Reactor Excursion Leak Analysis Program (RELAP5) thermal-hydraulic analyses, fracture mechanics analyses, and probabilistic risk assessment. This report documents the basis for the probabilistic fracture mechanics models used in the PTS reevaluation effort and encoded in the computer program FAVOR Version 04.1.

Brian W. Sheron, Director
Office of Nuclear Regulatory Research
U.S. Nuclear Regulatory Commission

CONTENTS

FIGURES

TABLES

EXECUTIVE SUMMARY

This report is one of a series of reports that summarize the results of a 5-year project conducted by the U.S. Nuclear Regulatory Commission's (NRC) Office of Nuclear Regulatory Research. This study sought to develop a technical basis to support revision of Title 10, Section 50.61, of the *Code of Federal Regulations* (10 CFR 50.61), which is known as the pressurized thermal shock (PTS) rule and the associated PTS screening criteria in a manner consistent with current NRC guidelines on risk-informed regulation. Figure ES-1 illustrates how this report fits into the overall project documentation.

This Executive Summary begins with a description of PTS, how it might occur, and what the potential consequences are for the vessel. A summary of the current regulatory approach to PTS follows, which leads directly to a discussion of the motivations for undertaking this project. This section concludes with a description of how the project was conducted. This introductory material provides a context for the information presented in this report concerning the details of the probabilistic fracture mechanics (PFM) model.

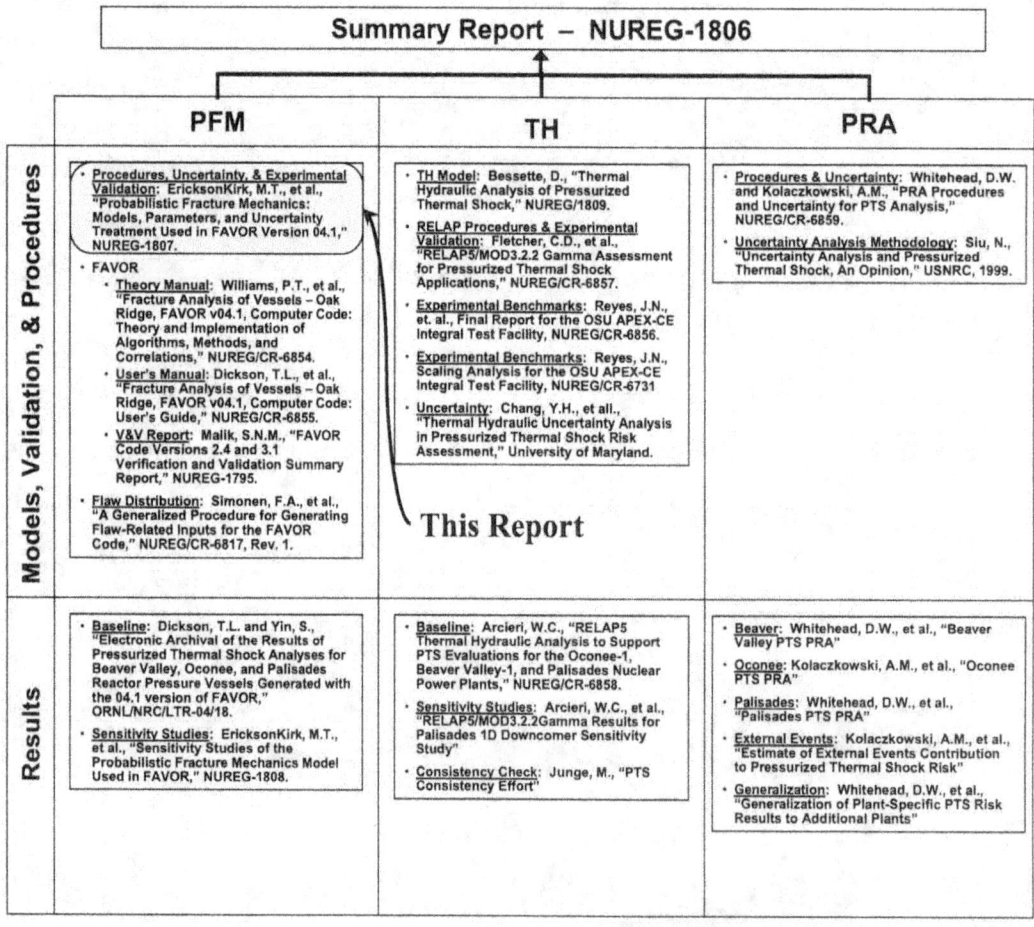

Figure ES-1 Structure of reports documenting the PTS reevaluation effort

ES.1 Description of PTS

One potentially significant challenge to the structural integrity of the reactor pressure vessel (RPV) in a pressurized-water reactor (PWR) is posed by a PTS event in which rapid cooling of the downcomer occurs, possibly followed by repressurization. A number of abnormal events and postulated accidents have the potential to thermally shock the vessel (either with or without significant internal pressure); some of these include a pipe break in the primary pressure circuit, a stuck-open valve in the primary pressure circuit, and the break of the main steamline. During these events, the water level drops because of the contraction produced by rapid depressurization. In events involving a break in the primary pressure circuit system, the water level drops further because of leakage from the break. Automatic systems and operators must provide makeup water in the primary system to prevent overheating of the fuel in the core. The makeup water is much colder than that held in the primary system.

The temperature drop produced by rapid depressurization, coupled with the near-ambient temperature of the makeup water, produces significant thermal stresses in the thick-section steel wall of the RPV. For embrittled RPVs, these stresses could be high enough to initiate a running crack that could propagate all the way through the vessel wall. Through-wall cracking of the RPV could precipitate core damage or, in rare cases, a large early release of radioactive material to the environment.

ES.2 Current PTS Regulations

As required by 10 CFR 50.61, licensees must monitor the embrittlement of their RPVs using a surveillance program qualified by Appendix H to 10 CFR Part 50. The results of surveillance are used together with the formulae and tables in 10 CFR 50.61 to estimate the fracture toughness transition temperature (RT_{NDT}) of the steels in the vessel's beltline, as well as how these transition temperatures increase because of irradiation damage throughout the operational life of the vessel. For licensing purposes, 10 CFR 50.61 provides instructions on how to use these estimates of the effect of irradiation damage on RT_{NDT} to estimate the value of RT_{NDT} that will occur at end of license (EOL), a value called RT_{PTS}. In addition, 10 CFR 50.61 provides "screening limits," or maximum values of RT_{NDT}, permitted during the operating life of the plant of +132 °C (+270 °F) for axial welds, plates, and forgings and +149 °C (+300 °F) for circumferential welds. These screening limits correspond to a limit of 5×10^{-6} events/yr on the yearly probability of developing a through-wall crack (see Regulatory Guide (RG) 1.154, "Format and Content of Plant-Specific Pressurized Thermal Shock Safety Analysis Reports for Pressurized Water Reactors"). Should RT_{PTS} exceed these screening limits, 10 CFR 50.61 requires that the licensee either take actions to keep it below the screening limit (i.e., by implementing "reasonably practicable" flux reductions to reduce the embrittlement rate or by deembrittling the vessel by annealing (see RG 1.162, "Thermal Annealing of Reactor Pressure Vessel Steels") or perform a plant-specific analysis to demonstrate that operating the plant beyond the 10 CFR 50.61 screening limit does not pose an undue risk to the public (see RG 1.154, "Format and Content of Plant-Specific Pressurized Thermal Shock Safety Analysis Reports for Pressurized Water Reactors").

While no currently operating PWR has an RT_{PTS} value that exceeds the 10 CFR 50.61 screening limits before EOL, several plants are close to the limit (3 are within 1°C (2 °F) while 10 are within 11 °C (20 °F)). Those plants that are close to the limit are likely to exceed it during the 20-year license renewal period that many operators are currently seeking. Moreover, some plants maintain their RT_{PTS} values below the 10 CFR 50.61 screening limits by implementing flux reduction (low-leakage cores; ultra-low leakage cores) and other fuel management strategies that can be economically deleterious in a deregulated marketplace. Thus, the 10 CFR 50.61 screening limits can restrict the licensable and the economic lifetime of PWRs.

ES.3 Motivation for This Project

It is now widely recognized that the state of knowledge and data limitations in the early 1980s necessitated a conservative treatment of several key parameters and models used in the probabilistic calculations that provide the technical basis of the current PTS rule. The most prominent of these conservatisms include the following:

- the highly simplified treatment of plant transients (i.e., the very coarse grouping of many operational sequences (on the order of 10^5) into very few groups (approximately 10)) necessitated by limitations in the computational resources needed to perform multiple thermal hydraulic (TH) calculations

- the lack of any significant credit for operator action

- the characterization of fracture toughness using RT_{NDT}, which has an intentional conservative bias

- the use of a flaw distribution that placed all of the flaws on the interior surface of the RPV, and, in general, contains larger flaws than those usually detected in service

- the modeling approach that treated the RPV as if it were made entirely from the most brittle of its constituent materials (welds, plates, or forgings)

- the modeling approach that assessed RPV embrittlement using the peak fluence over the entire interior surface of the RPV

These factors indicate the high likelihood that the current 10 CFR 50.61 PTS screening limits are unnecessarily conservative. Consequently, it was believed that a reexamination of the technical basis for these screening limits that is based on a modern understanding of all the factors that influence PTS would most likely strongly justify a substantial relaxation of these limits. For these reasons the NRC's Office of Nuclear Reactor Regulation undertook this project with the objective of developing the technical basis to support a risk-informed revision of the PTS rule and the associated PTS screening limit.

ES.4 Approach

As illustrated in Figure ES-2, there are three main models (shown as solid blue squares) that, together, allow an estimate of the yearly frequency of through-wall cracking in an RPV:

(1)   a probabilistic risk assessment event sequence analysis
(2)   a thermal hydraulic analysis
(3)   a probabilistic fracture mechanics analysis

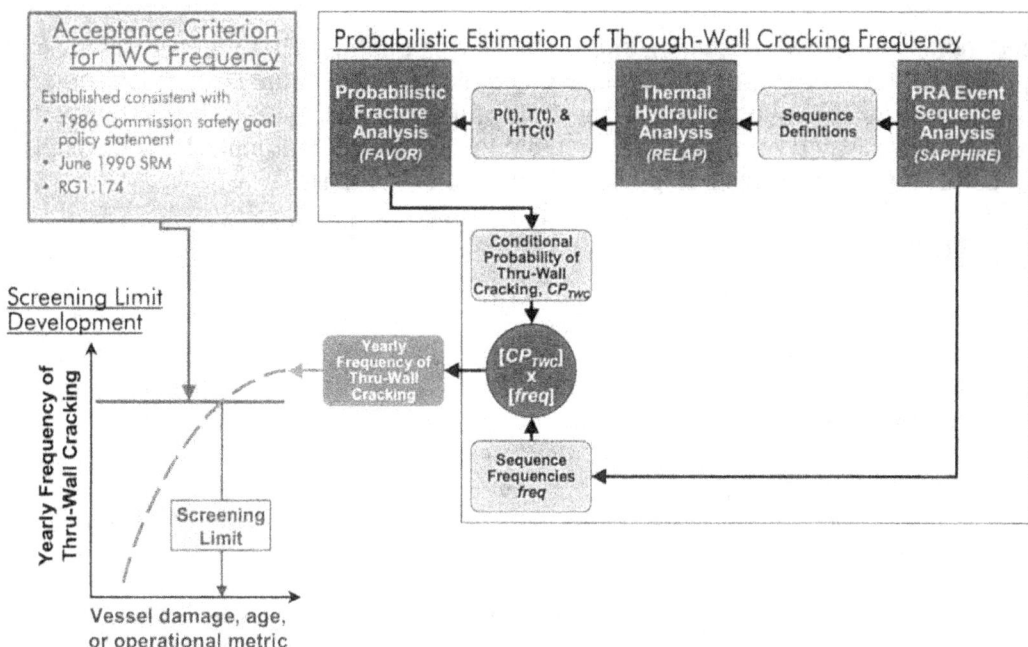

Figure ES-2 Schematic showing how a probabilistic estimate of TWCF is combined with a TWCF acceptance criterion to arrive at a proposed revision to the PTS screening limit

A probabilistic risk assessment (PRA) event sequence analysis is first performed to define the sequences of events that are likely to produce a PTS challenge to RPV integrity and to estimate the frequency with which such sequences can be expected to occur. The event sequence definitions are then passed to a TH model that estimates the temporal variation of temperature, pressure, and heat transfer coefficient in the RPV downcomer characteristic of each of the sequence definitions. These pressure, temperature, and heat transfer coefficient histories are passed to a probabilistic fracture mechanics (PFM) model, which uses the TH output, along with other information concerning plant design and materials of construction, to estimate the time-dependent driving force to fracture produced by a particular event sequence. The PFM model compares this estimate of fracture driving force to the fracture toughness, or fracture resistance, of the RPV steel. This comparison allows an estimate of the probability that a particular sequence of events will produce a crack all the way through the RPV wall, if that sequence of events were to actually occur. The final step in the analysis involves a simple matrix multiplication of the probability of through-wall cracking (from the PFM analysis) with the frequency at which a particular event sequence is expected to occur (as defined by the event-tree analysis). This product establishes an estimate of the yearly frequency of through-wall cracking that can be expected for a particular plant after a particular period of operation when subjected to a particular sequence of events. The yearly frequency of through-wall cracking is then summed for all event sequences to estimate the total yearly frequency of through-wall cracking for the vessel. Performance of such analyses for various operating lifetimes provides an estimate of how the yearly through-wall cracking frequency (TWCF) can be expected to vary over the lifetime of the plant.

The probabilistic calculations just described are performed to establish the technical basis for a revised PTS rule within an integrated systems analysis framework. The NRC approach considers a broad range of factors that influence the likelihood of vessel failure during a PTS event while accounting for uncertainties in these factors across a breadth of technical disciplines. Two central features of this approach are (1) a focus on the use of realistic input values and models (wherever possible), and (2) an explicit treatment of uncertainties (using currently available uncertainty analysis tools and techniques). Thus, the NRC's current approach improves upon that employed in the development of SECY-82-465, which included intentional and unquantified conservatisms in the many aspects of the analysis, and which treated uncertainties implicitly by incorporating them into the models.

ES.5 Key Findings

As discussed earlier, one of the technical motivations for this project is the understanding that the state of knowledge and data limitations in the early 1980s necessitated a conservative treatment of several key parameters and models used in the probabilistic calculations that provide the technical basis of the current PTS rule. Some of the most substantive conservatisms exist in the PFM model, which include the following:

- the characterization of fracture toughness using RT_{NDT} which has an intentional conservative bias

- the use of a flaw distribution that placed all of the flaws on the interior surface of the RPV, and that, in general, contained larger flaws than those usually detected in service

- the modeling approach that treated the RPV as if it were made entirely from the most brittle of its constituent materials (welds, plates, or forgings)

- the modeling approach that assessed RPV embrittlement using the peak fluence over the entire interior surface of the RPV

These and other conservatisms motivated the NRC to fundamentally reexamine and restructure the PFM model as this report summarizes. The major accomplishments of the improvements made to the PFM model are described in the following paragraphs.

- This report provides a thorough and systematic examination of all parts of the PFM model; it reveals many instances in which uncertainties were previously treated implicitly through the use of conservative models and parameter inputs. In the revised model (documented in this report) the NRC has, to the greatest extent possible, removed all such implicit treatments. Where conservative approaches are still part of the model (most often in areas in which adequate knowledge is lacking), they are explicitly identified.

- This review has also identified the nature of uncertainties in the models and parameters that collectively make up the PFM model (i.e., as being aleatory or epistemic) and to quantify them.

- The PFM model consists of a crack initiation model, an embrittlement model, and a through-wall cracking model. The major features of and improvements to the PFM model are as follows:

  o Crack Initiation Model
    - This model included the removal, on average, of the large conservative bias in the RT_{NDT} transition temperature. This was achieved by recognizing that RT_{NDT}

does *not* measure fracture toughness, and by adopting alternative metrics that do measure fracture toughness.
- It also included separation of uncertainties in the crack initiation model into its epistemic (resulting from of RTNDT) and its aleatory (resulting from the scatter inherent to ferritic steels) parts, allowing their separate and proper quantification.
  - o Embrittlement Model
    - Local variations in both fluence and material properties are recognized, an improvement over the old representation in which the vessel was modeled as being comprised entirely of the most radiation-sensitive material exposed to the highest fluence in the vessel.
    - The damaging effect of radiation on the fracture toughness of ferritic steels is represented using a model with a functional form based on an understanding of the physical mechanisms responsible for irradiation damage. As such, the ability of this model to extrapolate beyond the conditions for which it was calibrated is superior to models used previously, which were predominantly empirical in origin.
  - o Through-Wall Cracking Model
    - The through-wall cracking model recognizes the ability of ferritic steels to arrest cleavage crack propagation at high applied driving forces.
    - As radiation damage increases, the cleavage crack initiation toughness of ferritic steels will approach the cleavage crack arrest toughness. The through-wall cracking model now incorporates this feature, removing a feature from old models that was physically unrealistic.
    - The through-wall cracking model now recognizes and accounts for the possibility of the RPV failing by ductile (rather than cleavage) mechanisms. Past models assumed that ductile failure was impossible and, in so doing, introduced nonconservatisms into the model.

- The PFM model includes the following features which must be viewed as conservative:
  - o The model explicitly considers uncertainty in copper, in nickel, and in initial RT_{NDT}. However, it represents these uncertainties as being larger (a conservative representation) than would be appropriate in any plant-specific application.
  - o The model used to represent the attenuation or radiation damage through the thickness of the RPV is conservative (i.e., the model predicts an increase in toughness through wall (from inner diameter to outer diameter) that is smaller than that revealed by experiments).
  - o Once a circumferential crack initiates, the model assumes that it will instantly propagate 360° around the vessel wall. Full circumferential propagation is highly unlikely because of the azimuthal variation in fluence, which causes alternating regions of more embrittled and less embrittled material to exist circumferentially around the vessel wall. Thus, the NRC model tends to overestimate the extent of cracking initiated from circumferentially oriented defects because it ignores this natural crack arrest mechanism.
  - o Once an axial flaw initiates, the model assumes that it will instantly become infinitely long. In reality it will only propagate to the length of an axial shell course (approximately 8 to 12 feet), at which point it will encounter tougher material and arrest. Even though the length of a shell course is very long, finite length flaws tend to arrest more readily than do infinite length flaws because of systematic differences in the through-wall variation of the crack driving force. Because of this approximation, the NRC model tends to overestimate the likelihood of through-wall cracking.

o   The experimental data upon which the flaw distribution is based modeled all detected defects as being crack-like, and therefore potentially deleterious to the fracture integrity of the vessel. However, many of these defects are actually volumetric rather than planar, making them either benign or, at a minimum, much less of a challenge to the fracture integrity of the vessel. The NRC model thus overestimates the seriousness of the defect population in RPV materials, which leads to overly pessimistic assessments of the fracture resistance of the vessel.

1 INTRODUCTION

1.1 Description of Pressurized Thermal Shock

One potentially significant challenge to the structural integrity of the reactor pressure vessel (RPV) in a pressurized-water reactor (PWR) is posed by a pressurized thermal shock (PTS) event in which rapid cooling of the downcomer occurs, possibly followed by repressurization. A number of abnormal events and postulated accidents have the potential to thermally shock the vessel (either with or without significant internal pressure); some of these include a pipe break in the primary pressure circuit, a stuck-open valve in the primary pressure circuit, and the break of the main steamline. During these events, the water level drops because of the contraction produced by rapid depressurization. In events involving a break in the primary pressure circuit system, the water level drops further because of leakage from the break. Automatic systems and operators must provide makeup water in the primary system to prevent the fuel in the core from overheating. The makeup water is much colder than that held in the primary system.

The temperature drop produced by rapid depressurization, coupled with the near-ambient temperature of the makeup water, produces significant thermal stresses in the thick-section steel wall of the RPV. For embrittled RPVs, these stresses could be high enough to initiate a running crack that could propagate all the way through the vessel wall. Through-wall cracking of the RPV could precipitate core damage or, in rare cases, a large early release of radioactive material to the environment.

1.2 PTS Limits on the Licensable Life of a Commercial Pressurized-Water Reactor

In the early 1980s attention was focused on the possibility that PTS events could challenge the integrity of the RPV wall for two reasons:

(1)   Operational experience suggested that overcooling events, while not common, did in fact occur.

(2)   The results of in-reactor materials surveillance programs suggested that the steels used in RPV construction were prone to loss of toughness over time as the result of neutron irradiation-induced embrittlement.

This possibility of accident loading combined with degraded material conditions motivated investigations to assess the risk of vessel failure posed by PTS in order to establish the operational limits needed to ensure that the likelihood of RPV failures caused by PTS transients remained sufficiently low. These efforts led to the publication of a document (SECY-82-465) that provided the technical basis for subsequent development of what has come to be known as the "PTS rule" (Title 10, Section 50.61, of the *Code of Federal Regulations* (10 CFR 50.61)).

According to 10 CFR 50.61, licensees must monitor the embrittlement of their RPVs using a surveillance program qualified by Appendix H to 10 CFR Part 50. The results of surveillance are used together with the formulae and tables in 10 CFR 50.61 to estimate the fracture toughness transition temperature ($RT_{NDT}{}^{\dagger}$) of the steels in the vessel's beltline, as well as how these transition temperatures increase because of irradiation damage throughout the operational life of the vessel. For licensing purposes, 10 CFR 50.61 provides instructions on how to use such estimates of the effect of irradiation damage on RT_{NDT} to estimate the value of RT_{NDT} that will occur at end of license (EOL), a value called RT_{PTS}. In addition, 10 CFR 50.61 provides screening limits, or maximum values of RT_{NDT}, permitted during the operating life of the plant of +132 °C (+270 °F) for axial welds, plates, and forgings and +149 °C (+300 °F) for circumferential welds. These screening limits correspond to a limit of 5×10^{-6} events/yr on the yearly probability of developing a through-wall crack (see Regulatory Guide (RG) 1.154, "Format and Content of Plant-Specific Pressurized Thermal Shock Safety Analysis Reports for Pressurized Water Reactors"). Should RT_{PTS} exceed these screening limits, 10 CFR 50.61 requires that the licensee either take actions to keep it below the screening limit (i.e., by implementing "reasonably practicable" flux reductions to reduce the embrittlement rate or by de-embrittling the vessel by annealing (see RG 1.162, "Thermal Annealing of Reactor Pressure Vessel Steels") or perform a plant-specific analysis to demonstrate that operating the plant beyond the 10 CFR 50.61 screening limit does not pose an undue risk to the public (see RG 1.154, "Format and Content of Plant-Specific Pressurized Thermal Shock Safety Analysis Reports for Pressurized Water Reactors").

While no currently operating PWR has an RT_{PTS} value that exceeds the 10 CFR 50.61 screening limit before EOL, several plants are close to the limit (3 are within 1 °C (2 °F) while 10 are within 11 °C (20 °F); see Figure 1-1). Those plants that are close to the limit are likely to exceed it during the 20-year license renewal period that many operators are currently seeking. Moreover, some plants maintain their RT_{PTS} values below the 10 CFR 50.61 screening limits by implementing flux reduction (low-leakage cores; ultra-low leakage cores) and other fuel management strategies that can be economically deleterious in a deregulated marketplace. Thus, the 10 CFR 50.61 screening limits can restrict the licensable and the economic lifetime of PWRs. As detailed in the next section, there is considerable reason to believe that these restrictions are not necessary to ensure public safety and, in fact, place an unnecessary burden on licensees.

1.3 Technical Factors Suggesting Conservatism of the Current Rule

It is now widely recognized that the state of knowledge and data limitations in the early 1980s necessitated a conservative treatment of several key parameters and models used in the probabilistic calculations that provide the technical basis (see SECY-82-465) of the current PTS rule (10 CFR 50.61). The most prominent of these conservatisms include the following:

- the highly simplified treatment of plant transients (i.e., the very coarse grouping of many operational sequences (on the order of 10^5) into very few groups (approximately 10)) necessitated by limitations in the computational resources needed to perform multiple thermal hydraulic calculations

- the lack of any significant credit for operator action

---

† The RT_{NDT} index temperature was intended to correlate with the fracture toughness transition temperature of the material. Fracture toughness, and how it is reduced by neutron irradiation embrittlement, is a key parameter controlling the resistance of the RPV to any loading challenge. For a more detailed description of RT_{NDT} (in specific) and fracture toughness (in general) see EricksonKirk 10-03.

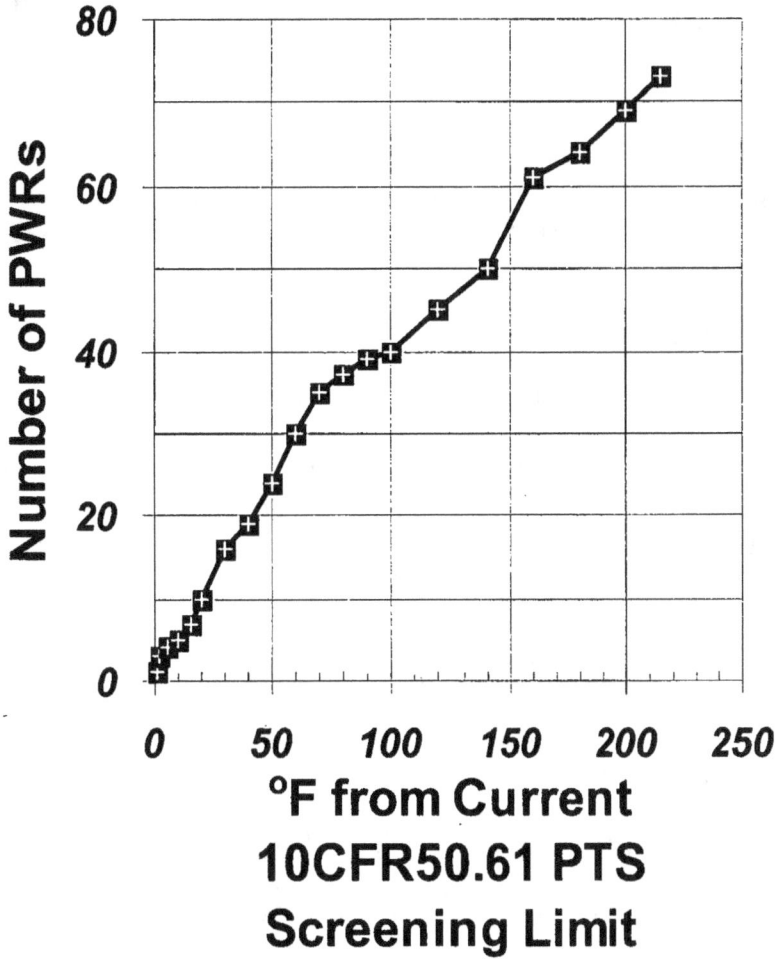

Figure 1-1 Proximity of currently operating PWRs to the 10 CFR 50.61 screening limit for PTS

- the characterization of fracture toughness using RT_{NDT} which has an intentional conservative bias (see ASME NB2331)

- the use of a flaw distribution that placed all of the flaws on the interior surface of the RPV and, in general, contained larger flaws than those usually detected in service

- the modeling approach that treated the RPV as if it were made entirely from the most brittle of its constituent materials (welds, plates, or forgings)

- the modeling approach that assessed RPV embrittlement using the peak fluence over the entire interior surface of the RPV

These factors indicate the high likelihood that the current 10 CFR 50.61 PTS screening limits are unnecessarily conservative. Consequently, a new examination of the technical basis for these screening limits based on a modern understanding of all the factors that influence PTS would most likely strongly justify a substantial relaxation of these limits. For these reasons, the U.S. Nuclear Regulatory Commission's (NRC) Office of Nuclear Regulatory Research undertook this project with the objective of developing the technical basis to support revision of the PTS rule and the associated PTS screening limit.

1.4 PTS Reevaluation Project

This section describes the PTS reevaluation project, which the NRC Office of Nuclear Regulatory Research initiated in 1999. It discusses restrictions placed on the model used to estimate PTS risk, the overall structure of the model, how the model addressed uncertainties, and how the this and other reports document the results of the project.

1.4.1 Restrictions on the Model

The desired outcome of this research effort is the establishment of the technical basis for a new PTS screening limit. To enable all commercial operators of PWRs to assess the state of their RPV relative to such a new criterion, without the need to make new material property measurements, the fracture toughness properties of the RPV steels need to be estimated using only information that is currently available (i.e., RT_{NDT} values, upper-shelf energy values, and the chemical composition of the beltline materials). All of this information is summarized in the NRC Reactor Vessel Integrity Database (RVID2).

1.4.2 Overall Structure of the Model

The NRC's overall model involves three major components, which are illustrated (along with their interactions), in Figure 1-2:

(1)   The first component, <u>probabilistic evaluation of through-wall cracking frequency (TWCF), involves estimating</u> the frequency of through-wall cracking as a result of a PTS event given the operating, design, and material conditions in a particular plant.

(2)   The second component, <u>acceptance criterion for TWCF, involves establishing</u> a value of reactor vessel failure frequency (RVFF) consistent with current guidance on risk-informed decisionmaking.

(3)   The third component, <u>screening limit development</u>, involves comparing the results of the two preceding steps to determine if some simple, materials-based screening limit for PTS can be established. Conceptually, plants falling below the screening limit would be deemed adequately resistant to a PTS challenge and would not require further analysis. Conversely, more detailed, plant-specific analysis would be needed to assess the safety of a plant's operation beyond the screening limit.

The following subsections describe each of these components.

1.4.2.1   Component 1—Probabilistic Estimation of Through-Wall Cracking Frequency

As illustrated in Figure 1-2, there are three main models (shown as solid blue squares) that together allow an estimate of the yearly frequency of through-wall cracking in an RPV:

(1)   a probabilistic risk assessment (PRA) event sequence analysis
(2)   a TH analysis
(3)   a probabilistic fracture mechanics (PFM) analysis

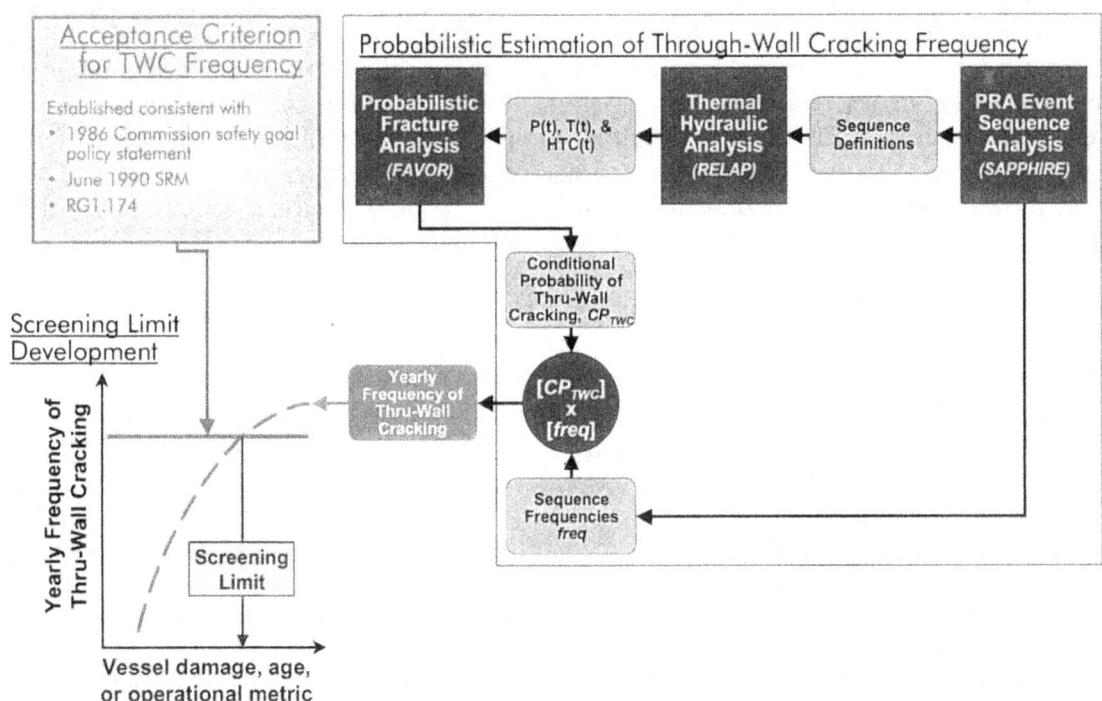

Figure 1-2 High-level schematic showing how a probabilistic estimate of TWCF is combined with a TWCF acceptance criterion to arrive at a proposed revision to the PTS screening limit

The following subsections first describe these three models in general and then describe their sequential execution to provide the reader with an appreciation for the interrelationships and interfaces between the different models (Section 1.4.2.1.1). Next, the subsections describe the iterative process the NRC undertook, which involved repeated execution of all three models in sequence, to arrive at final models for each plant (Section 1.4.2.1.2). Finally, the subsections discuss the three specific plants the NRC analyzed in detail (Section 1.4.2.1.3). This section concludes with a discussion of the steps taken to ensure that the NRC's conclusions based on these three analyses apply to domestic PWRs in general (Section 1.4.2.1.4).

1.4.2.1.1 Sequential Description of How PRA, TH, and PFM Models *Are* Used to Estimate TWCF

A PRA event sequence analysis is first performed to define the sequences of events that are likely to produce a PTS challenge to RPV integrity and to estimate the frequency with which such sequences can be expected to occur. The event sequence definitions are then passed to a TH model that estimates the temporal variation of temperature, pressure, and heat transfer coefficient in the RPV downcomer characteristic of each of the sequence definitions. These pressure, temperature, and heat transfer coefficient histories are passed to a PFM model, which uses the TH output, along with other information concerning plant design and materials of construction, to estimate the time-dependent driving force to fracture produced by a particular event sequence. The PFM model compares this estimate of fracture driving force to the fracture toughness, or fracture resistance, of the RPV steel. This comparison allows an estimate of the probability that a particular sequence of events will produce a crack all the way through the RPV wall if that sequence of events were to actually occur. The final step in the analysis involves a simple matrix multiplication of the probability of through-wall cracking (from the PFM analysis) with the frequency at which a particular event sequence is expected to occur (as defined by the event-tree analysis). This product establishes an estimate of the yearly frequency of through-wall cracking that can be expected for a particular plant after a particular period of operation when subjected to a particular sequence of events. The yearly frequency of through-wall cracking is then summed for all event sequences to estimate the total yearly frequency of through-wall cracking for the vessel. Performance of such analyses for various operating lifetimes provides an estimate of how the yearly TWCF can be expected to vary over the lifetime of the plant.

1.4.2.1.2 Iterative Process Used to Establish Plant-Specific Models

The set of transients used to represent a particular plant are identified using a PRA event-tree approach, in which many thousands of different initiating event sequences are "binned" together into groups of transients believed to produce similar TH outcomes. Judgments regarding what transients to put into what bin were guided by such characteristics as similarity of break size and similarity of operator action, resulting in bins such as "medium break primary system loss-of-coolant accidents (LOCAs)" and "main steamline breaks". From each of the tens or hundreds of individual event sequences in each bin, the NRC selected a single sequence and programmed it into the TH code RELAP to define the variation of pressure, temperature, and heat transfer coefficient *vs.* time. These TH transient definitions were then passed to the PFM code FAVOR, which estimated the conditional probability of through-wall cracking (CPTWC) for each transient. When multiplied by the initiating event frequency estimates estimated in the PRA analysis, these CPTWC become TWCF values, which, when rank-ordered, estimate the degree to which each bin contributes to the total TWCF of the vessel. At this stage many bins are found to contribute very little or nothing at all to the TWCF, and so receive little further scrutiny. However, some bins invariably dominate the TWCF estimate. These bins are then further subdivided by partitioning the initiating event frequency of the bin, and by selecting a TH transient to represent each part of the original bin. FAVOR is then used to analyze this refined model,, and the bins that provide significant contributions to TWCF are again examined. This process of bin partitioning, and the selection of a TH transient to represent each newly partitioned bin, continues until the total estimated TWCF for the plant no longer changes significantly.

1.4.2.1.3 Plant-Specific Analyses Performed

In this project, the NRC performed detailed calculations for three operating PWRs (Oconee Unit 1, Beaver Valley Unit 1, and Palisades; see Figure 1-3). Together the three plants sample a wide range of design and construction methods, and they contain some of the most embrittled RPVs in the current operating fleet.

1.4.2.1.4 Generalization to all Domestic PWRs

Because the objective of this project is to develop a revision to the PTS screening limit expressed in 10 CFR 50.61 that applies in general to all PWRs, the NRC must understand to what extent these three plant-specific analyses adequately address (in either a representative or in a bounding sense) the range of conditions experienced by domestic PWRs in general. The NRC performed the following actions to achieve this goal:

- High embrittlement plant
- Westinghouse design

- High embrittlement plant
- Combustion Engineering design

- Plant used in 1980s PTS study
- Babcox & Wilcox design

Figure 1-3 The three plants analyzed in detail in the PTS reevaluation effort

- The NRC performed sensitivity studies on both the TH and PFM models to address the effect of credible changes to the model and/or its input parameters. The results of these studies provide insights regarding how robust the NRC's conclusions on the three plants are when applied to the PWR population in general.

- The NRC examined the plant design and operational characteristics of five additional plants. The aim of this additional analysis was to identify if the design and operational features identified as being important in the three plant-specific analyses vary significantly enough in the general population to question the generality of the results.

- In the three plant-specific analyses, the NRC assumed that the only possible causes of PTS events have origins that are internal to the plant. However, external events, such as fires, floods, and earthquakes, can also be PTS precursors. The NRC therefore examined the potential for external initiating events to create significant additional risk relative to the internal initiating events already modeled in detail.

1.4.2.2   Component 2—Acceptance Criterion for Through-Wall Cracking Frequency

Since the issuance of SECY-82-465 and the publication of the original PTS rule, the NRC has established a considerable amount of guidance on the use of risk metrics and risk information in regulation (e.g., the Safety Goal Policy Statement, the PRA Policy Statement, and RG 1.174, "An Approach for Using Probabilistic Risk Assessment in Risk-Informed Decisions On Plant-Specific Changes to the Licensing Basis"). To ensure the consistency of the PTS reevaluation project with this guidance, the staff identified and assessed options for a risk-informed criterion for the RVFF (currently specified in RG 1.154 in terms of TWCF).

As described in SECY-02-0092, the options developed involve both qualitative concerns (the definition of RPV failure) and quantitative concerns (a numerical criterion for the RVFF). The options reflected uncertainties in the margin between PTS-induced RPV failure, core damage, and large early release. The options also incorporated input received from the Advisory Committee on Reactor Safeguards (ACRS) (see NRC LTR 02) regarding concerns over the potential for large-scale oxidation of reactor fuel in an air environment.

The NRC's assessment of the options involved the identification of technical issues unique to the PTS accident scenario development, the development of an accident progression event tree to structure consideration of the issues, the performance of a scoping study of the issue of containment performance during PTS accidents, and the review of the options in light of this information. The scoping study involved collecting and evaluating available information, performing a few limited-scope TH and structural calculations, and a semiquantitative analysis of the likelihood of various accident progression scenarios.

1.4.2.3   Component 3—Screening Limit Development

As illustrated schematically in Figure 1-2 (lower left corner), a screening limit for PTS can be established based on a simple comparison of estimates of the RVFF as a function of an appropriate measure of RPV embrittlement with the RVFF acceptance criterion (or RVFF*). Beyond the work needed to establish both the RVFF vs. embrittlement curve and RVFF* values, it is also necessary to establish a suitable vessel damage metric that, ideally, allows different conditions in different materials at different plants to be normalized. From a practical standpoint, "suitable" implies that the metric needs to be based only on information regarding plant operation and materials that is readily available.

1.4.3   Uncertainty Treatment

At the outset of this project in 1999, the NRC staff reviewed the Agency's existing approach for PRA modeling, focusing on how uncertainties should be treated, how they were propagated through the PRA, TH, and PFM models, and how that approach compared with the NRC's guidelines on work supporting risk-informed regulation (see Siu 99). This review established a general framework for model development and uncertainty treatment, which the following paragraphs summarize.

This project performed probabilistic calculations to establish the technical basis for a revised PTS rule within an integrated systems analysis framework (see Woods 01). The NRC approach considers a broad range of factors that influence the likelihood of vessel failure during a PTS event while accounting for uncertainties in these factors across a breadth of technical disciplines (see Siu 99). Two central features of this approach are (1) a focus on the use of realistic input values and models (wherever possible), and (2) an explicit treatment of uncertainties (using currently available uncertainty analysis tools and techniques). Thus, the current approach improves upon that employed in the development of SECY-82-465, which included intentional and unquantified conservatisms in the many aspects of the analysis, and which treated uncertainties implicitly by incorporating them into the models (RT_{NDT}, for example).

The NRC's probabilistic models distinguish between two types of uncertainties, aleatory and epistemic. Aleatory uncertainties result from the randomness inherent to a physical or human process, whereas epistemic uncertainties are caused by a limitation in the current state of knowledge (or understanding) of that process. A practical way to distinguish between aleatory and epistemic uncertainties is that epistemic uncertainties can, in principle, be reduced by an increased state of knowledge. Conversely, because aleatory uncertainties result from randomness at a level below which a particular process is modeled, they are fundamentally irreducible. The distinction between aleatory and epistemic uncertainties is an important part of the PTS analysis because different mathematical and/or modeling procedures are used to represent these different uncertainty types.

1.4.4   Project Documentation

This report is one of a series of reports that summarize the results of a PTS reevaluation project. Figure 1-4 illustrates how this report fits into the overall structure of the project documentation.

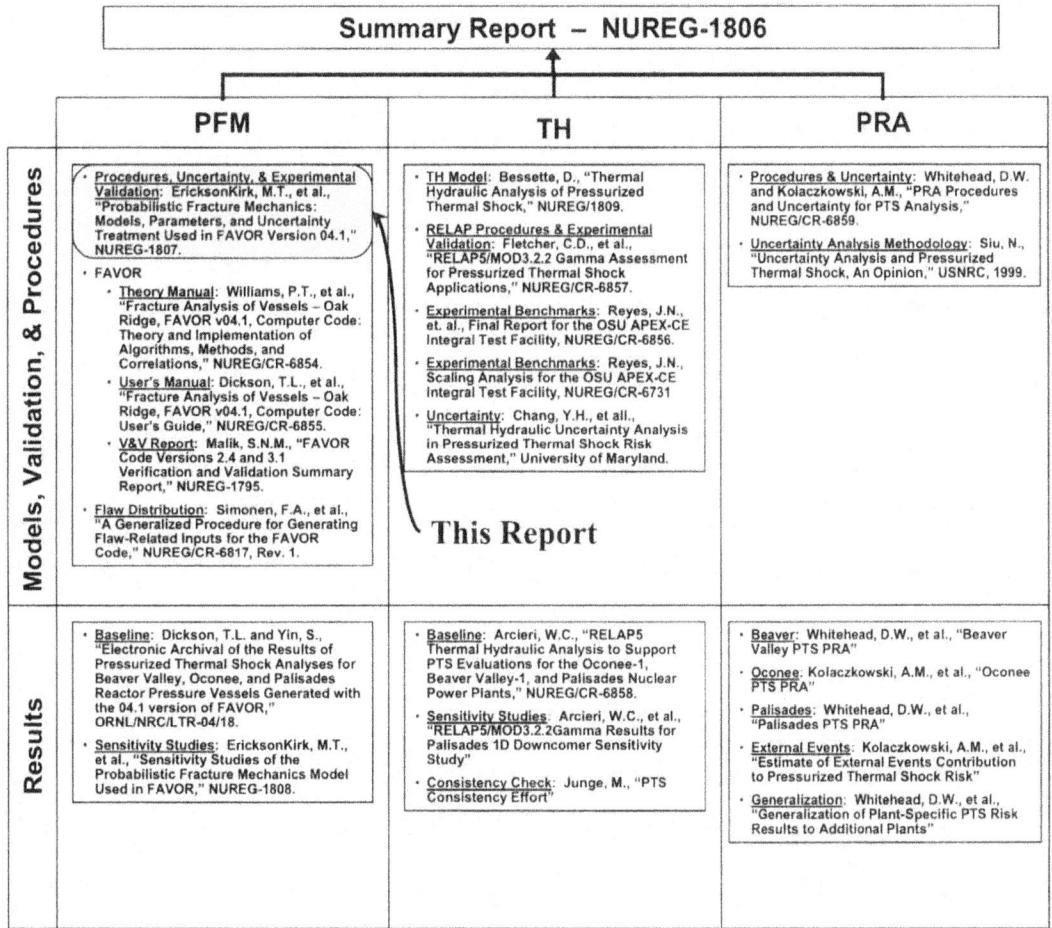

Figure 1-4 Structure of reports documenting the PTS reevaluation effort

2 OBJECTIVE, SCOPE, AND STRUCTURE OF THIS REPORT

This report describes the models and the parameters that make up the probabilistic fracture mechanics (PFM) model that has been implemented in FAVOR Version 04.1 (see Williams 04). Additionally, this report describes how uncertainties in these models and parameter inputs are treated. Figure 1-2 illustrates where PFM fits, together with a probabilistic risk assessment (PRA) and a thermal hydraulic (TH) analysis, within the overall model used to estimate the through-wall cracking frequency (and its uncertainty). Figure 2-1 provides a more detailed depiction of the interaction of the PFM model (large shaded boxes) with the PRA and TH models (unshaded boxes), and of the PFM model itself.

The PFM model contains the following three main parts:

(1)   a flaw distribution model (the uppermost of the large shaded boxes in Figure 2-1)
(2)   a crack initiation model (the middle of the large shaded boxes in Figure 2-1)
(3)   a through-wall cracking model (the lowest of the large shaded boxes in Figure 2-1)

As illustrated in Figure 2-1, each of these models is itself a complex assemblage of submodels and parameter inputs. The flaw distribution model and the treatment of uncertainties in the flaw distribution model is the subject of a companion report (see Simonen 03) and will not be discussed herein.

This report begins with a description of the fundamental assumptions underlying the U.S. Nuclear Regulatory Commission's (NRC) modeling approach (see Chapter 3). A detailed description of the crack initiation model (with uncertainty treatment) and of the through-wall cracking model (with uncertainty treatment) follows; see Chapters 4 and 5, respectively.

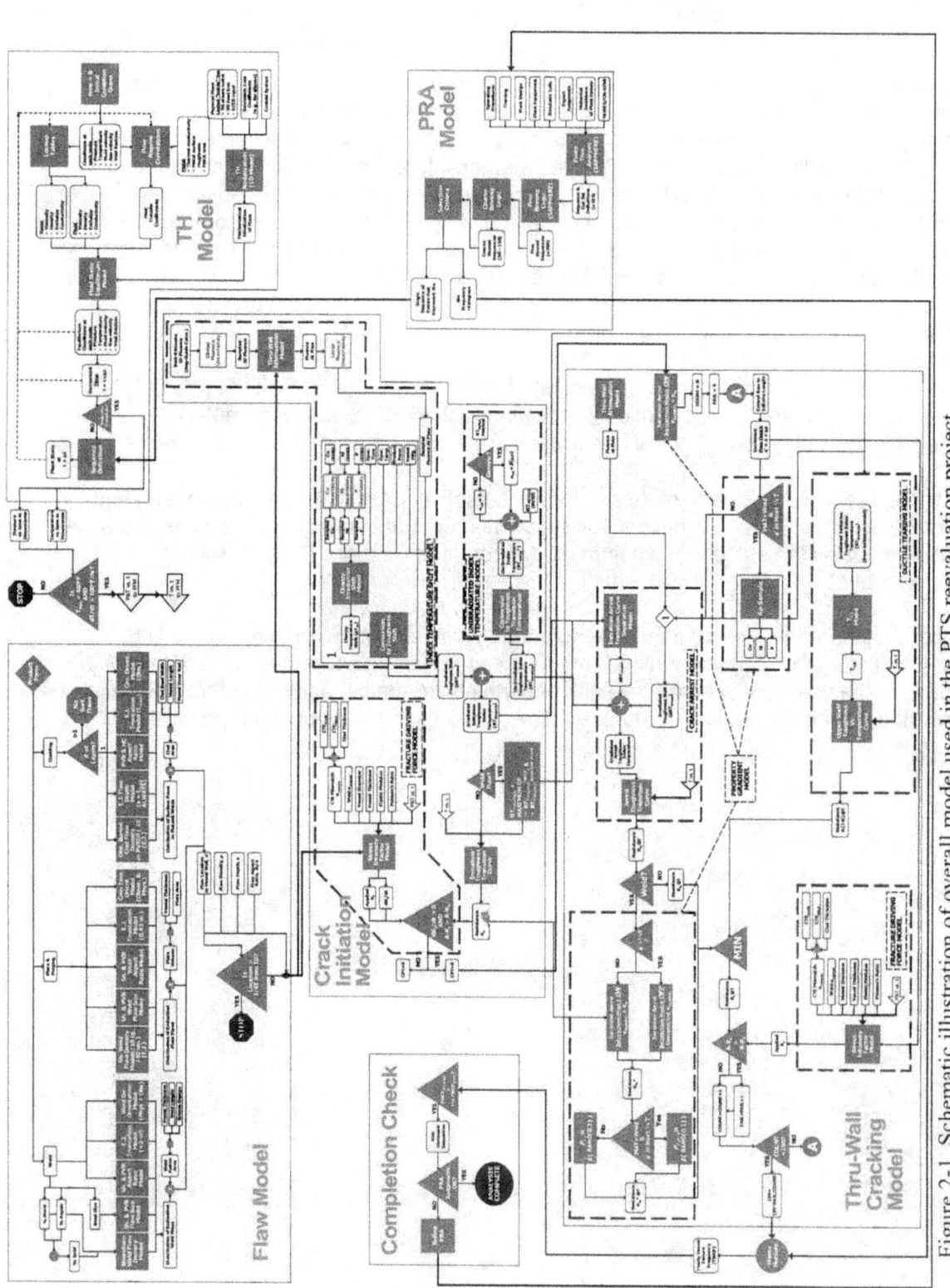

Figure 2-1 Schematic illustration of overall model used in the PTS reevaluation project

2-2

3 FUNDAMENTAL ASSUMPTIONS

The appropriateness of the probabilistic fracture mechanics (PFM) analysis performed by FAVOR to assess pressurized thermal shock (PTS) rests on the validity of the following four fundamental assumptions:

(1)   The U.S. Nuclear Regulatory Commission (NRC) assumes, in general, that linear elastic fracture mechanics (LEFM) is an appropriate methodology to use in assessing the structural integrity of reactor pressure vessels (RPVs) subjected to PTS loadings, and, specifically, that FAVOR predictions of the fracture response of RPVs in response to PTS loading are accurate.

(2)   The NRC assumes that the effects of crack growth by subcritical mechanisms (i.e., environmentally assisted cracking and/or fatigue) is negligible, and consequently the flaw population of interest is that associated with initial vessel fabrication.

(3)   The NRC assumes that the fracture toughness of the stainless steel cladding is adequately high, and remains adequately high even after irradiation, that there is no possibility of failure of the cladding due to the loading imposed by PTS transients.

(4)   The NRC assumes that stresses occur in sufficiently low locations in the vessel wall (between $3/8 \cdot t_{wall}$ from the vessel inside diameter and the outside diameter) that the probability of failure associated with postulated defects in this region does not need to be calculated because it is zero.

(5)   The NRC assumes that if a particular transient does not achieve a temperature in the downcomer below 204 °C (400 °F), then it does not contribute to the vessel failure probability.

The following subsections discuss the appropriateness of each of these assumptions.

3.1   Use of Linear Elastic Fracture Mechanics

One fundamental assumption the NRC made in constructing the PFM model is that a linear elastic stress analysis of the vessel, and a consequent fracture integrity assessment using the techniques of LEFM, is accurate. Evidence supporting the appropriateness of this assumption is available in the following two areas:

(1)   Appendix A summarizes the results of studies aimed at experimentally validating the appropriateness of LEFM techniques when assessing the integrity of nuclear pressure vessels under thermal shock and PTS experiments. The results of three series of experiments that Oak Ridge National Laboratory performed in the 1970s and 1980s on scaled pressure vessels demonstrate the accuracy of LEFM techniques in these applications.

(2)   One of the fundamental requirements for LEFM "validity" is that the dimensions of the plastic zone at the tip of a loaded crack be very small as compared to the dimensions of the crack being assessed and the dimensions of the structure in which the crack resides (see Rolfe). Under these conditions, the errors introduced by plastic flow (which is not accounted for within LEFM theories) are acceptably small. To assess plastic zone sizes characteristic of the PTS problem, the NRC had the PFM code FAVOR report all of the applied driving force to fracture ($K_{APPLIED}$) values from an analysis of Beaver Valley Unit 1 at 60 effective full-power years (EFPYs) that contribute to the through-wall cracking frequency (TWCF) (i.e., that have a conditional probability of crack initiation greater than 0). The top graph in Figure 3-1 shows these $K_{APPLIED}$ values overlaid on the K_{Ic} transition curve, while the bottom figure shows these same values expressed in the form of a cumulative distribution function. The lower figure indicates that 90 percent of the $K_{APPLIED}$ values that contribute to the TWCF estimate lie between 22-39 MPa√m (20-35 ksi√in). Using these stress intensity factor values, together with Irwin's equation for the plastic zone size under plane-strain conditions (see Rolfe), indicates that the plastic zone radii characteristic of PTS loading range from approximately 0.8-3.3 mm (0.03-0.13 in.), depending on the value of $K_{APPLIED}$ (here taken to range from 22-39 MPa√m (20-35 ksi√in)) and the value of the yield strengths (here taken to be 483 MPa (70 ksi) on average for unirradiated materials and 621 MPa (90 ksi) on average for irradiated materials). These values of plastic zone radii are certainly small as compared to the thickness of a pressurized-water reactor (PWR) vessel, indicating the appropriateness of LEFM techniques. Moreover, it can be noted that as the vessel ages, irradiation damage causes the yield strength to increase. Thus, as vessels approach end of life (EOL) and extended EOL conditions, LEFM techniques become, if anything, more appropriate.

3.2 Assumption of No Subcritical Crack Growth

3.2.1   Caused by Environmental Mechanisms Acting on the Low Alloy Pressure Vessel Steel

Stress-corrosion cracking (SCC) requires the presence of an aggressive environment, a susceptible material, and a significant tensile stress. If these three requirements are met and SCC can occur, the growth of intrinsic surface flaws in a material is possible. Because an accurate PTS calculation for the low-alloy steel (LAS) pressure vessel should address realistic flaw sizes, in principle, the potential for crack growth in the reactor vessel resulting from SCC needs to be analyzed. However, for the reasons detailed the following paragraphs, SCC of LAS in PWR environments is highly unlikely and, therefore, appropriately assumed not to occur for the purposes of the FAVOR calculations reported herein.

The first line of defense against SCC of LAS is the cladding that covers much of the LAS surface area of the reactor vessel and main coolant lines. This prevents the environment from contacting the LAS, and therefore obviates any possibility of SCC of the pressure boundary. Additionally, several test programs have been conducted over the last three decades, all of which show that in normal PWR or boiling-water reactor (BWR) operating environments, SCC in LAS cannot occur. The electrochemical potential (often called the free corrosion potential) controls SCC of LAS in the reactor coolant environment. The oxygen concentration in the coolant is the main variable that controls the LAS electrochemical potential. During normal operation of a PWR, the oxygen concentration is below 5 parts per billion (ppb). The electrochemical potential of LAS in this environment would not reach the value necessary to cause SCC (see IAEA 90, Hurst 85, Rippstein 89, and Congleton 85). During refueling conditions, the oxygen concentration in the reactor coolant does increase. However, the temperature during an outage is low, rendering SCC kinetically unfavorable. During refueling outage conditions with higher oxygen concentrations but lower temperatures, the electrochemical potential of the LAS would still not reach the values necessary for SCC to occur (see Congleton 85).

3.2.2 Caused by Environmental Mechanisms Acting on the Austenitic Stainless Steel Cladding

As stated in Section 3.2.1 one of the assurances of the negligible effects of environmentally assisted crack growth on the low alloy pressure vessel steel is the integrity of the austenitic stainless steel cladding that provides a corrosion resistant barrier between the LAS and the primary system water. Under conditions of normal operation the chemistry of the water in the primary pressure circuit is controlled with the express purpose of ensuring that stress corrosion cracking of the stainless steel cladding cannot occur. Even under chemical upset conditions (during which control of water chemistry is temporarily lost) the rate of crack growth in the cladding is exceedingly small. For example, Ruther et al. report an upper bound crack growth rate of $\approx 10^{-5}$ mm/s ($\approx 4 \times 10^{-7}$ in/s) in poor quality water (i.e., high oxygen) environments [Ruther 84]. The amount of crack extension that could occur during a chemical upset would is therefore quite limited, certainly not sufficient to compromise the integrity of the clad layer.

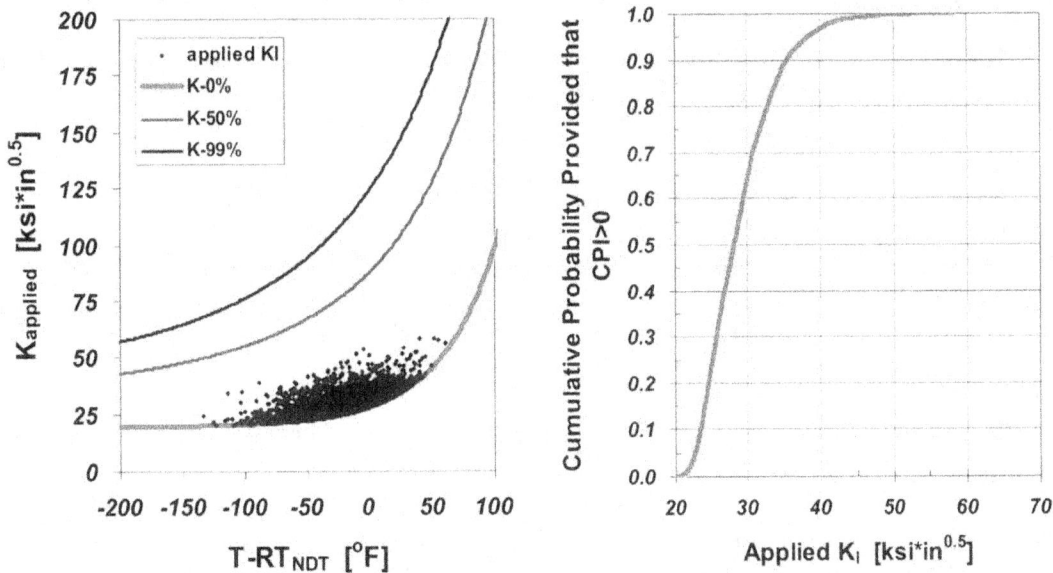

Figure 3-1 Illustration of the magnitude of $K_{APPLIED}$ values that contribute to the TWCF because they have a CPI > 0. Top figure shows all $K_{APPLIED}$ values with CPI > 0 overlaid on the K_{Ic} transition curve from an analysis of Beaver Valley Unit 1 at 60 EFPYs. Bottom figure shows these same results expressed in the form of a cumulative distribution function.

3.2.3 Caused by Fatigue

Fatigue is a mechanism that initiates and propagates flaws under the influence of fluctuating or cyclic applied stress and can be separated into two broad stages, (1) fatigue damage accumulation, potentially leading to crack initiation, and (2) fatigue crack growth.

Fatigue is influenced by variables that include mean stress, stress range, environmental conditions, and surface roughness and temperature. Thermal fatigue can also occur as thermal stresses develop when a material is heated or cooled. Generally, fatigue failures occur at stresses having a maximum value less than the yield strength of the material. The process of fatigue damage accumulation, crack initiation, and crack growth is closely related to the phenomenon of slip resulting from static shear stress. Following a period of fatigue damage accumulation, crack initiation will occur by the progressive development and linking up of intrusions along slip bands or grain boundaries. Growth of these initiated cracks includes fracture deformation sequences, plastic blunting followed by resharpening of the crack tip, and alternate slip processes.

The PWR vessel is specifically designed so that all of its components satisfy the fatigue design requirements of the American Society of Mechanical Engineer (ASME) Code Section III, or equivalent. Several studies have shown that the 60-year anticipated fatigue "usage" of the vessel beltline region resulting from normal plant operations, including plant heatup/cooldown and design-basis transients, is low, such that fatigue-initiated cracks will not occur. Similarly, fatigue loading of the vessel is considered insufficient to result in propagation of any existing fabrication defects (see EPRI 94, Kasza 96, and Kahleel 00).

3.3 Assumption that the Stainless Steel Cladding will not Fail as a Result of the Loads Applied by PTS

Stainless steel, even in the clad form, typically exhibits initiation fracture resistance (J_{Ic} and J-R) values that far exceed those of the ferritic steels from which the wall of the RPV is made (see [Bass 04] for cladding data, and see Section 5.2.2 for ferritic steel data). This is especially true for the levels of embrittlement at which vessel failure becomes a (small) probability because, at the fluences characteristic of the vessel inner diameter location, the fracture toughness of ferritic steels can be considerably degraded by neutron damage while the fracture toughness of austenitic stainless steels are essentially unaffected by these same levels of irradiation damage [Chopra 05]. This high toughness of the stainless steel cladding coupled with the small characteristic size of defects found in the cladding [Simonen 04] justifies the assumption that the stainless steel cladding will not fail as a result of the loads applied by PTS.

3.4 Noncontribution of Flaws Deep in the Vessel Wall to Vessel Failure Probability

The FAVOR flaws simulated to exist further than $3/8 \cdot t_{WALL}$ from the inner diameter surface are eliminated, *a priori*, from further analysis. This screening criterion is justified based on deterministic fracture mechanics analyses which demonstrate that for the embrittlement and loading conditions characteristic of PTS, such flaws have zero probability of crack initiation. As illustrated in Figure 3-2, in practice crack initiation almost always occurs from flaws that having their inner crack tip located within $0.2 \cdot t_{WALL}$ of the inner diameter, further substantiating the appropriateness of eliminating cracks deeper than $3/8 \cdot t_{WALL}$ from further analysis.

3.5 Noncontribution of Certain Transients to Vessel Failure Probability

When running a plant-specific analysis using FAVOR, the NRC only calculates the conditional probability of through-wall cracking (CPTWC) for thermal hydraulic (TH) transients that reach a minimum temperature at or below 204 °C (400 °F). Experience and deterministic calculations justify this *a priori* elimination of transients, and demonstrate that such transients lack adequate severity to have nonzero values of CPTWC, even for very large flaws and for very large degrees of embrittlement. Additionally, the results of the NRC's plant-specific analyses (see Chapter 8 of EricksonKirk 04a) show that a minimum transient temperature of 178 °C (352 °F) must be reached before CPTWC rises above zero, validating that the elimination of transients with minimum temperatures above 204 °C (400 °F) does not influence the results in any way.

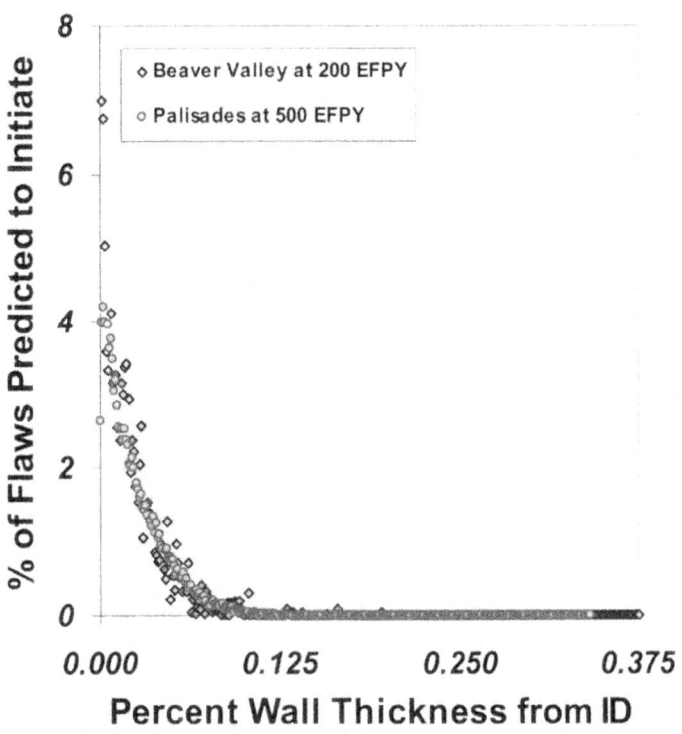

Figure 3-2 Distribution of crack initiating depths generated by FAVOR Version 3.1

4   CRACK INITIATION MODEL

The crack initiation model detailed in Figure 4-1 compares the applied driving force to fracture ($K_{APPLIED}$, shown in the shaded section of the figure) and the material's resistance to crack initiation in the cleavage (K_{Ic}, shown in the unshaded section of the figure). A comparison of $K_{APPLIED}$ (a single value at each time during the transient) and K_{Ic} (a distribution of values at each time in the transient) allows one to estimate if any conditional probability of crack initiation (CPI) exists (i.e., when $K_{APPLIED}$ is greater than or equal to K_{Ic}) or if no probability of crack initiation whatsoever exists (i.e., when $K_{APPLIED}$ is less than K_{Ic}).

Sections 4.1 and 4.2, respectively, address the component parts of the $K_{APPLIED}$ and K_{Ic} models, as well as the treatment of uncertainty in each model.

4.1   Applied Driving Force to Fracture

As illustrated in Figure 4-2, the model of applied driving force to fracture is a conventional linear elastic fracture mechanics (LEFM) driving force model (backed in light purple) augmented by a warm prestress (WPS) model (backed in light yellow). FAVOR implements both the LEFM and the WPS models deterministically. Sections 4.1.1 and 4.1.2, respectively, outline the rationale for adopting LEFM and WPS and for their deterministic treatment.

4.1.1   LEFM Driving Force

4.1.1.1   Appropriateness of the Model

Appendix A provides a detailed report describing why the use of an LEFM model is appropriate when assessing the risk of brittle failure caused by a pressurized thermal shock (PTS) event. This section provides a brief overview of this information.

Appendix A summarizes the findings from three extensive experimental/analytical investigations that examined the accuracy with which LEFM models could be expected to predict the failure of nuclear reactor pressure vessels (RPVs) subjected to both simple loadings (pressure only) and to much more complex loadings (PTS conditions). These investigations all featured tests on thick-section pressure vessels (see Figure 4-3), and aimed to reproduce, as closely as practical in a laboratory setting, the conditions that characterize thermal shock of a nuclear RPV. These conditions include the following:

- fracture initiation from small flaws

- severe thermal, stress, and material toughness gradients

- biaxial loading

- the effects of cladding (including residual stresses)

- conditions under which WPS may be active

4-1

- combined stress and toughness gradient conditions that can promote crack initiation, arrest, reinitiation, and rearrest all during the same transient

- the possibility of the conversion of the fracture mode from cleavage to ductile and back again all during the same thermal hydraulic (TH) transient, resulting from these various gradients

The three test series were as follows:

(1)   The first series of tests were 10 intermediate test vessels (ITVs), 3 with cracks located at a cylindrical nozzle and 7 with cracks remote from any geometric discontinuities. These tests were aimed at investigating the ability of LEFM to predict the fracture response of thick-section vessels containing relatively deep flaws (20–83 percent of the 6-in. thick vessel wall) at test temperatures ranging from lower shelf to upper shelf. Tests included a variety of nuclear grade RPV plates, forgings, and weldments.

(2)   The second series of tests were eight thermal shock experiments (TSEs). These experiments investigated the behavior of surface cracks under thermal shock conditions similar to those that would be encountered during a large-break loss-of-coolant accident (LBLOCA) (i.e., a rapid cooldown in the absence of internal pressure).

(3)   The third series of tests included two experiments that subjected ITV specimens to concurrent pressure and thermal transients. These pressurized thermal shock experiments (PTSEs) simulated the effects of a rapid cooldown transient combined with significant internal pressure. Thus, these experiments simulated TH conditions characteristic of smaller break loss-of-coolant accidents (LOCAs).

These investigations support the following conclusions:

- ITV experiments

  —   The LEFM analyses very closely predicted actual fracture pressures for thick-wall pressure vessels.

  —   Methods for calculating fracture toughness from small specimens were successfully used in applications of fracture analysis of thick, flawed vessels.

- thermal shock experiments

  —   Multiple initiation-arrest events with deep penetration into the vessel wall were predicted and observed.

  —   Surface flaws that were initially short and shallow were predicted and observed to grow considerably in length before increasing significantly in depth.

  —   WPS limited crack extension through the wall under LOCA conditions.

  —   Small-specimen fracture mechanics data successfully predicted the fracture behavior of thick pressure vessels

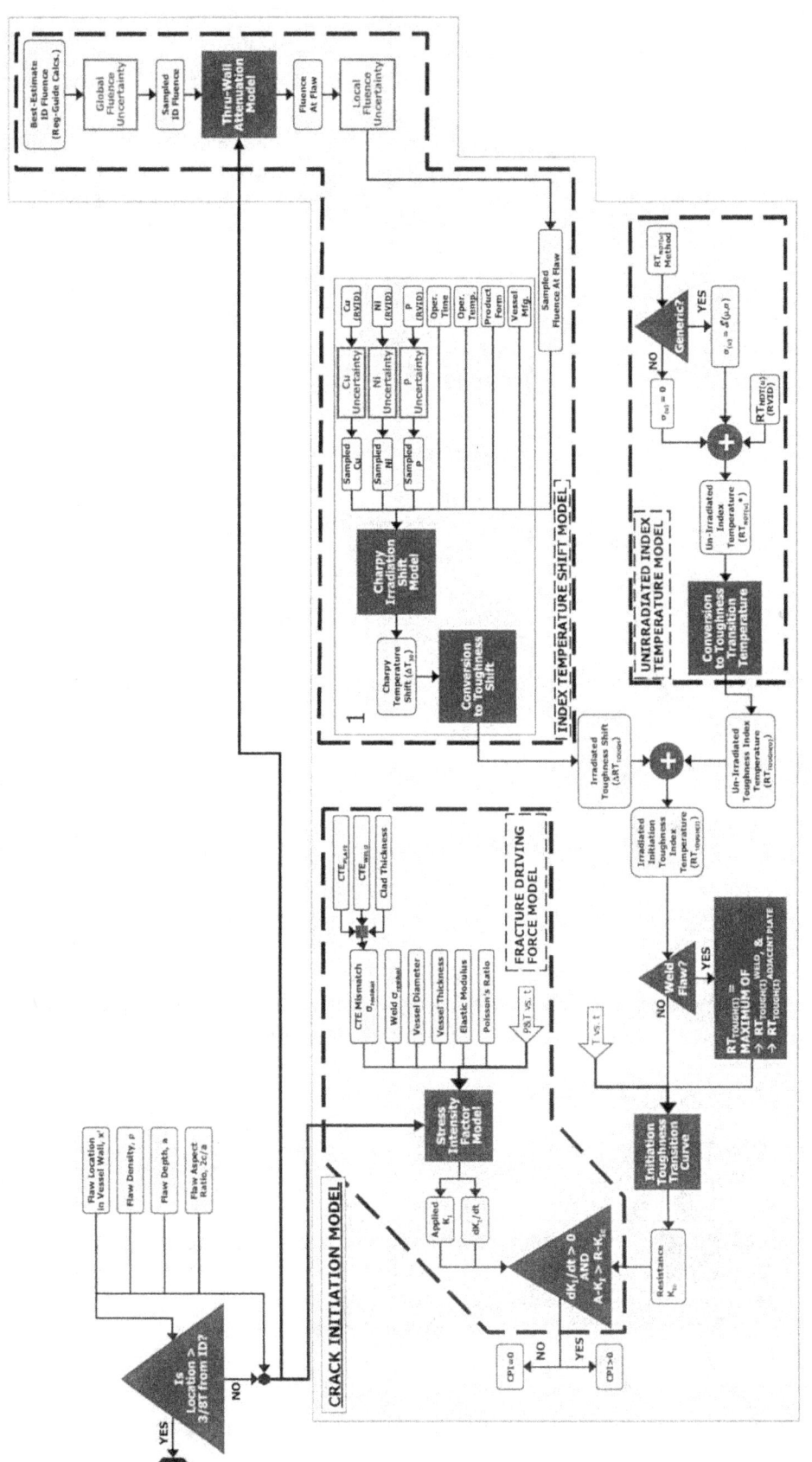

Figure 4-1 Schematic illustration of the crack initiation model used in the PTS reevaluation project

4-3

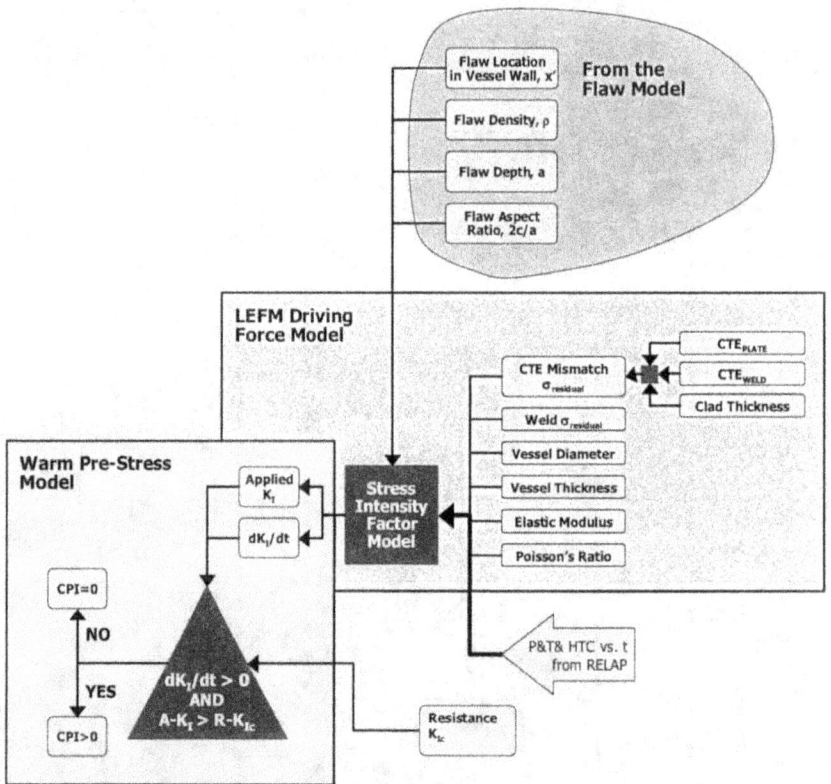

Figure 4-2 Schematic illustration of the model of the applied driving force to fracture used in the PTS reevaluation project

—   Crack arrest occurred in a rising stress field.

- pressurized thermal shock experiments

—   WPS is effective at inhibiting crack initiation for conditions under which crack initiation would otherwise be expected (i.e., $K_{APPLIED}$ is greater than K_{Ic}).

—   Crack arrest toughness values (K_{Ia}) inferred from conditions prototypic of PTS loading agree well with other experimental measurements, suggesting the transferability of laboratory toughness data to structural loading conditions.

—   LEFM predictions of crack initiation, growth, and arrest behavior successfully captured the response of the vessel to the transient; however, some details were not exactly predicted (for example: two initiation-run-arrest events were predicted whereas one was observed).

With regard to this final item, it should be noted that exact agreement between deterministic predictions and individual experiments cannot be expected when the physical processes that underlie those experiments produce large aleatory uncertainties (as is the case with K_{Ic} and K_{Ia} data; see Sections 4.2.2.3.2 and 5.1.2.2). Such disagreement does not in itself condemn the methodology, but rather reveals that the precision of any single prediction is limited by the precision in the knowledge of the controlling material properties.

4.1.1.2   Appropriateness of a Deterministic Implementation of LEFM in FAVOR

FAVOR deterministically modeled all of the material and geometric input variables to the LEFM model illustrated in Figure 4-2. In all cases, the deterministic input value represents a best estimate. The uncertainty in these parameters is very small (on the order of 10–20 percent of the mean value) relative to many other variables in the model that have their uncertainties modeled explicitly (K_{Ic}, for example, exhibits uncertainty on the order of the mean value, and the uncertainty in the initiating event frequency can be several orders of magnitude). In the face of these much larger uncertainties, it is not expected that the uncertainties of these input parameters influence the results of the computations significantly.

Figure 4-3 Test vessels used in the ITV and PTSE test series (top) and in the TSE test series (bottom)

The TH model RELAP estimates the pressure, temperature, and heat transfer coefficient inputs to the LEFM model illustrated by the arrow in Figure 4-2 (see RELAP 99 and RELAP 01). FAVOR all treats these inputs deterministically. This approach is appropriate because uncertainties in these TH inputs have already been addressed as part of the RELAP modeling process and in the way the probabilistic risk assessment (PRA) model represents bins of initiating event sequences using a single TH transient drawn from the bin.

4.1.2   Warm Prestress

4.1.2.1   Appropriateness of the Model

Appendix B contains a detailed discussion of the WPS phenomena and the appropriateness of accounting for WPS effects in PTS models. The information presented in this section summarizes that in Appendix B.

WPS effects were first noted in the literature in 1963 (see Brothers 63). These investigators reported (as have many since them) that the apparent fracture toughness of a ferritic steel can be elevated in the fracture mode transition if a fracture toughness test specimen is first "prestressed" at an elevated temperature. Once a specimen is subjected to a certain $K_{APPLIED}$ and has not failed, the temperature can be reduced, and the specimen will remain intact despite the fact that the process of reducing the temperature has also reduced the initiation fracture toughness to values smaller than $K_{APPLIED}$. In the past four decades, the technical community has identified, researched thoroughly, and agreed upon the physical mechanisms responsible for WPS.

As illustrated in Figure 4-4, $K_{APPLIED}$ first increases and then decreases as LOCA transients proceed, with the time of peak $K_{APPLIED}$ varying, depending on both the severity of the transient and the location of the crack in the vessel wall. It is the latter part of the transient when $K_{APPLIED}$ decreases with time that is of interest in the context of WPS. If the $K_{APPLIED}$ value generated by a LOCA were to enter the temperature-dependent distribution of initiation fracture toughness values during the falling portion of the transient, then the WPS phenomenon suggests that crack initiation will not occur even though $K_{APPLIED}$ exceeds the initiation fracture toughness of the material (see Figure 4-5).

In the past, probabilistic calculations performed in the United States to assess the challenge to RPV integrity posed by PTS events have not included WPS as part of the probabilistic fracture mechanics (PFM) model (see SECY-82-465, ORNL 85a, ORNL 85b, and ORNL 86), in spite of broad consensus in the technical community that WPS is a real phenomena having a sound physical basis. Idealizations in both the TH and PRA models prompted the exclusion of WPS from PFM models. As a consequence of these idealizations it was possible that the models could incorrectly represent a situation when WPS would not occur as a situation in which it could occur, which is a non-conservative error. The information in Appendix B demonstrates that the much more detailed PRA and TH models adopted as part of this PTS reevaluation effort eliminate this concern, now making inclusion of WPS appropriate.

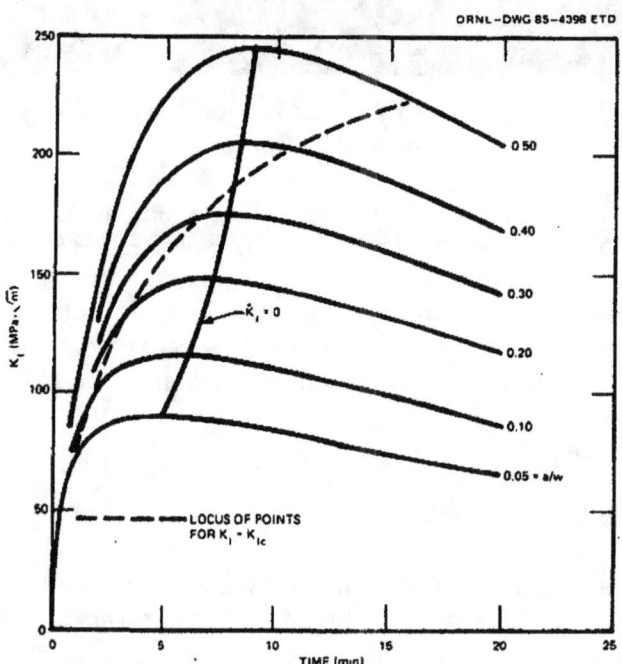

Figure 4-4 Illustration of the influence of crack depth on the variation of $K_{APPLIED}$ vs. time resulting from an LBLOCA (Cheverton 85)

4.1.2.2   Appropriateness of a Deterministic Implementation of WPS in FAVOR

Factors affecting the WPS model include only the crack driving force ($K_{APPLIED}$) and the fracture toughness (K_{Ic}). These models, and the appropriate treatment of uncertainty in each, appear elsewhere in this report (see Sections 4.1.1 and 4.2). An independent treatment of uncertainty in the WPS model is therefore not necessary.

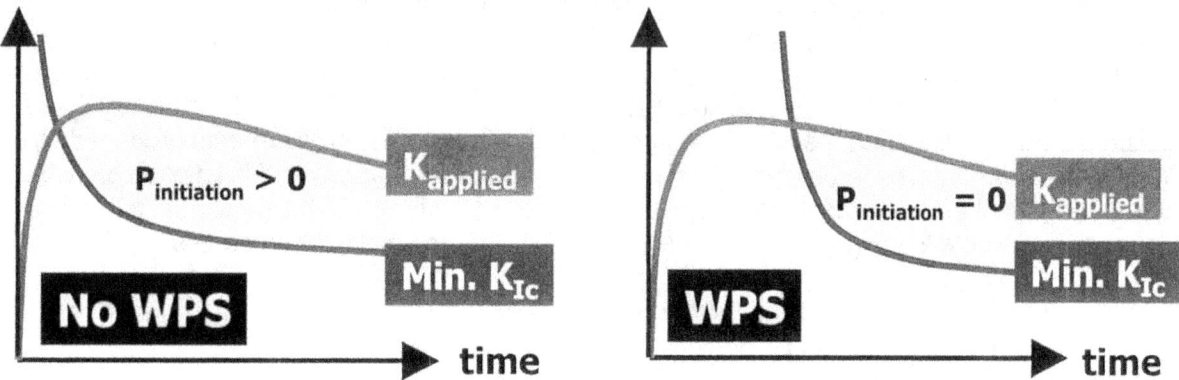

Figure 4-5 Schematic diagram illustrating how the WPS effect could be active during a LOCA depending on the combination of the transient and the position of the crack within the vessel wall

4.2   Resistance to Crack Initiation in Cleavage

As illustrated in Figure 4-6, the model of the resistance of a ferritic RPV steel to crack initiation in cleavage includes four major components:

(1)   an unirradiated index temperature model
(2)   a toughness transition model
(3)   an index temperature shift model
(4)   an interface model

As suggested by these names, the idea of using an index temperature approach to the characterization of ferritic steel fracture toughness and irradiation damage is central to the U.S. Nuclear Regulatory Commission's (NRC) computational approach. Therefore, Section 4.2.1 presents the evidence supporting an index temperature characterization. Section 4.2.2 uses this information to construct both the unirradiated index temperature model and the toughness transition model. This is followed by Section 4.2.3, which addresses the index temperature shift model, and by Section 4.2.4, which addresses the interface model.

4.2.1 Index Temperature Approaches to Characterizing the Transition Fracture Toughness of Ferritic Steels

The use of index temperature approaches to characterize the transition fracture (cleavage fracture toughness) properties of ferritic steels pervades the literature, dating back to the late 1940s. Qualitative uses of index temperature approaches derive from the observation of a temperature at which the steel transitions from brittle behavior (at lower temperatures) to ductile behavior (at higher temperatures). These approaches have been applied to characterize the fracture performance of both laboratory specimens (e.g., the Charpy V-notch (CVN) specimen (see ASTM E23), the nil ductility transition (NDT) specimen (see ASTM E208)), as well as full-scale structures (e.g., liberty ships (see Pellini 76)).

More recently, index temperatures have been incorporated into characterizations of fracture toughness, such as the ASME K_{Ic} and K_{Ia} curves, which use the fracture toughness transition temperature (RT_{NDT}) as an index temperature, and the "Master Curve" proposed by Wallin, which uses T_o as an index temperature (see Wallin All). The use of an index-temperature approach to characterize the fracture toughness of ferritic steels assumes (1) both the temperature dependency of fracture toughness and the scatter in fracture toughness at any one temperature are features that are common to a very broad class of materials (in this situation all RPV steels, all product forms, and all irradiation conditions of interest), and (2) provided these assumptions are true, then the index temperature alone reflects all of the effects produced by steel-to-steel differences in composition, heat treatment, product form, and irradiation damage. Strong empirical and physical evidence, as discussed in Sections 4.2.1.1 (universal temperature dependency) and 4.2.1.2 (universal scatter), demonstrate the validity of these assumptions.

4.2.1.1   Basis for a Universal Temperature Dependency of Fracture Toughness

4.2.1.1.1 Initiation Fracture Toughness

Over the past 5 years, Natishan and others (see Natishan All and Kirk 01a) have demonstrated that a physical basis for a temperature dependency common to all ferritic steels in fracture-mode transition can be found in dislocation mechanics. Below is an overview of this physical basis, as well as supporting empirical evidence.

Dislocation motion through the crystal occurs as atoms change position relative to each other, or "jump" between equilibrium lattice sites, as shown in Figure 4-7. This motion is opposed by a friction (named Peierls-Nabarro) stress produced by the presence of other atoms in the lattice. For dislocation "jumping" to occur, there must be enough energy supplied to the system, either by an externally applied mechanical stress or by thermal energy, to enable dislocations to overcome these short range barriers and change position. This process results in plastic flow of the material. The amount of energy required for dislocation motion through these short-range barriers depends on atom spacing within the lattice, and on the amplitude of atom vibrations about their lattice positions (which depends on temperature). At temperatures above absolute zero, atoms vibrate about their lattice positions because of the thermal energy in the system. As temperature increases, the amplitude of atom vibrations increases, resulting in an increased probability that an atom at any particular lattice site will be "out of position" at any given time. As atoms move out of position, the activation energy for dislocation motion around them is reduced. This lower activation energy reduces the applied shear stress required for dislocation motion and, thus, for plastic deformation. This effect manifests at the macroscale as a temperature dependency of both strength and toughness properties.

The discussion in the preceding paragraph suggests that the physical feature of steels that is responsible for the temperature dependency of properties is the short-range barriers to dislocation motion established by the lattice structure (which is body centered cubic, or BCC). Further consideration reveals that the lattice structure is the _only_ physical variable responsible for the temperature dependency of properties. Other factors usually thought to distinguish between "different" steels include the composition, thermomechanical treatment (i.e., product form), and degree of irradiation. These differentiating factors influence only those microstructural features having large interbarrier spacings (many tens or hundreds of atoms) relative to the atomic scale associated with the lattice structure (i.e., grain size/boundaries, point

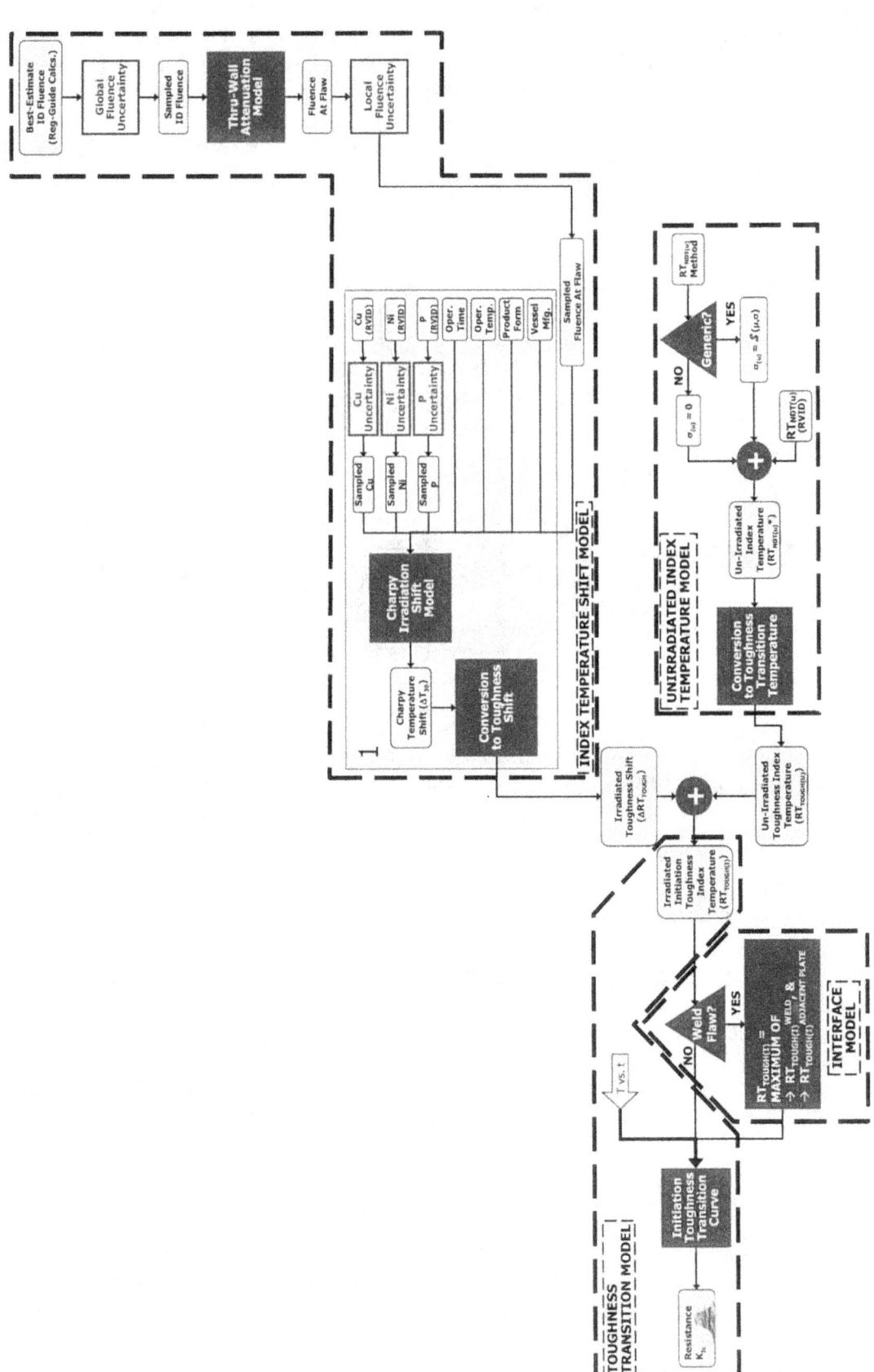

Figure 4-6   Schematic illustration of the model of the crack initiation toughness used in the PTS reevaluation project

4-9

defects, inclusions, precipitates, and dislocation substructures). Factors that produce only changes in large interbarrier spacings cannot influence on the temperature dependency of mechanical properties because the magnitude of lattice atom vibration that is influenced by temperature is not sufficient to affect the ease with which dislocations can travel around barriers having interatomic spacings larger than the atoms themselves. Thus, the myriad of factors normally thought of as differentiating steels from each other (e.g., composition, thermomechanical treatment, and irradiation) are not expected to have any influence on the temperature dependency of toughness in fracture mode transition; instead, they are seen to affect only the transition index temperature.

The work of Natishan et al. summarized in the preceding paragraphs demonstrates physical a basis for the expectation that all BCC materials should exhibit an identical temperature dependency, and that factors such as chemical composition, thermomechanical processing (product form), and level of irradiation damage should have no influence on this item. Empirical assessments employing large databases of both RPV steels and other ferritic steels (see Wallin 89, Sokolov 96, and Kirk 98) have validated these expectations. By way of example, Figure 4-8 and Figure 4-9 demonstrate the insignificance of both product form and radiation damage level in establishing the temperature dependency of fracture toughness, in agreement with the theoretical basis put forth by Natishan and co-workers.

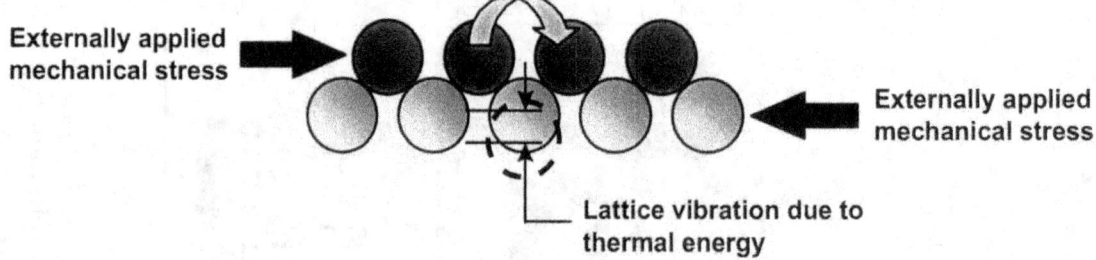

Figure 4-7 Illustration of the combined effects of mechanical stress and thermal energy on the ability of atoms to jump between equilibrium lattice sites

4.2.1.1.2 Arrest Fracture Toughness

Crack arrest occurs when dislocations can move faster than the crack propagates, which causes crack tip blunting and, thereby, arrest. Dislocation mobility therefore controls the ability of a ferritic steel to arrest a running cleavage crack, and thus its crack arrest toughness. The atomic lattice structure is the only feature of the material that controls the temperature-dependence of the material properties that are controlled by dislocation motion. Consequently, as was the case for crack initiation toughness, the temperature dependency of crack arrest toughness depends only on the short-range barriers to dislocation motion established by the BCC lattice structure. Other features that vary with steel composition, heat treatment, and irradiation include grain size/boundaries, point defects, inclusions, precipitates, and dislocation substructures. These features all influence dislocation motion, and thereby both strength and toughness, but their large interbarrier spacing relative to the atomic scale associated with the lattice structure makes these effects completely athermal. This understanding suggests that the myriad of metallurgical factors that can influence absolute strength and toughness values, and thereby the transition temperature, exert no control over the temperature dependency of arrest toughness in fracture mode transition. Additionally, since K_{Ic} and K_{Ia} both depend on the ability of the material to absorb energy by means of dislocation motion, K_{Ic} and K_{Ia} are both expected to exhibit a similar temperature dependence (see Kirk 02a). As was the case with crack initiation toughness, available empirical evidence demonstrates that the crack arrest toughness

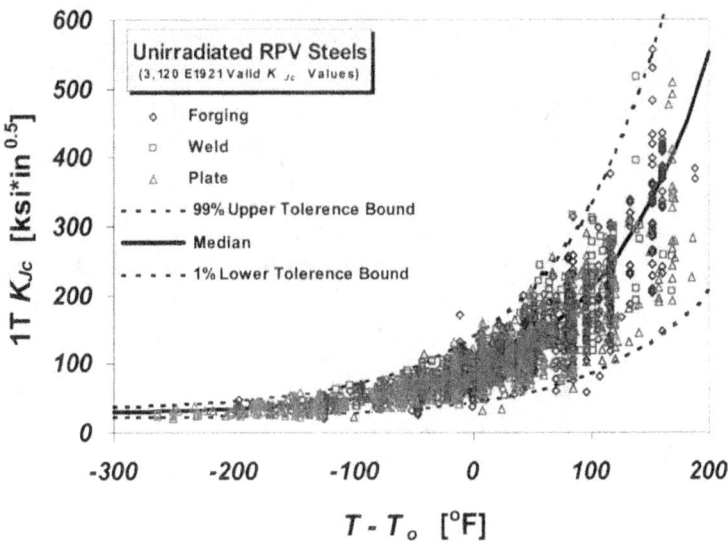

Figure 4-8 Illustration of the effect of product form on transition fracture toughness temperature dependency

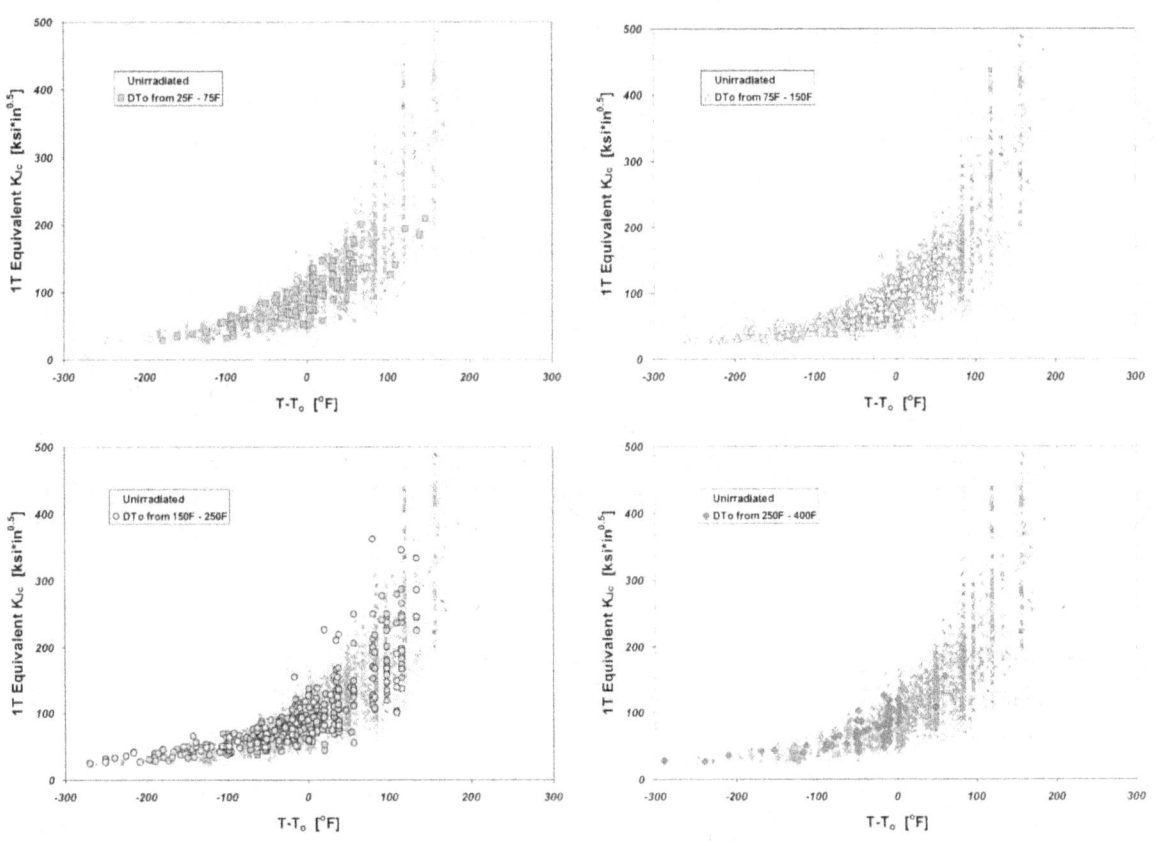

Figure 4-9 Illustration of the effect of radiation damage level on transition fracture toughness temperature dependency

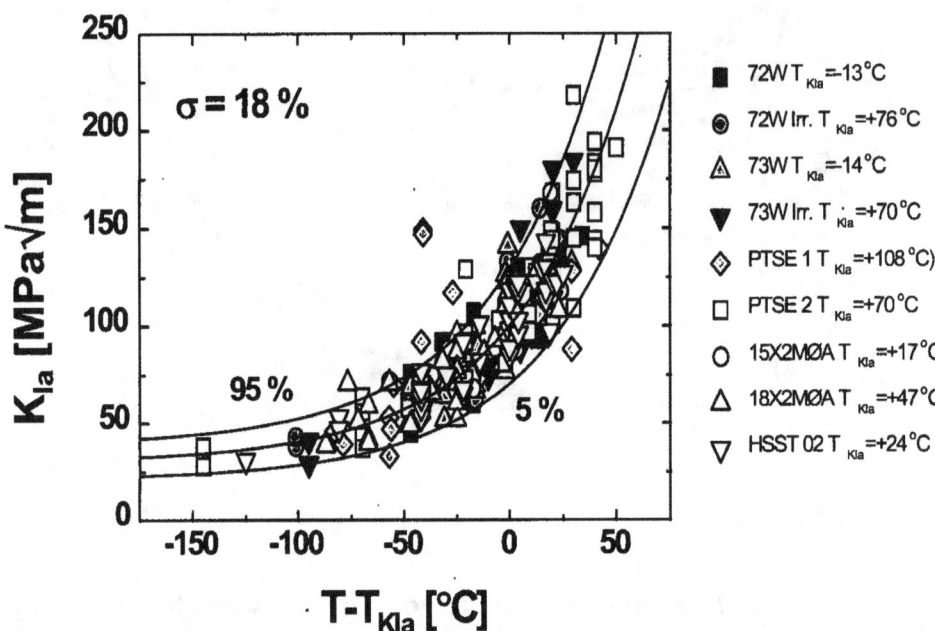

Figure 4-10 Crack arrest transition curves for nine heats of RPV steel. The mean curve has the same temperature dependence as the Master Curve for crack initiation data (i.e.,

$$K_{Ia} = 30 + 70 \cdot \exp\{0.019[T - T_{KIa}]\}$$ **(Wallin 97))**

of many different ferritic steels can be described by a common temperature dependency (see Wallin 97), as shown in Figure 4-10.

4.2.1.2   Basis for a Universal Scatter in Fracture Toughness

4.2.1.2.1   Initiation Fracture Toughness

Cleavage cracks initiate when the dislocations accumulated at noncoherent particles or other barriers to dislocation motion (e.g., carbides, grain boundaries, and twin boundaries) generate sufficient strain to elevate the local stress at the barrier above that needed to fracture the barrier or cause its decohesion from the matrix. These barriers are distributed in a random fashion throughout the BCC iron lattice. The interaction of these randomly distributed barriers with the varying stress field along the crack front gives rise to the experimentally observed scatter in toughness data.

In order for fracture to occur by cleavage, high stress triaxiality is necessary to inhibit crack-tip blunting by dislocation motion. Thus, for cleavage to occur, the stress fields must be in a state of small-scale yielding (SSY). High triaxiality occurs under SSY conditions because the crack-tip stress field is not affected by the specimen boundaries. This means that dislocations are fully contained within a finite volume at the crack tip and cannot escape to blunt the crack or dissipate energy. Under SSY conditions, the volume in which dislocations are moving can be described relative only to the length $L = (K_I/\sigma_y)^2$, making the total volume of the plastic fracture process zone proportional to $L^2 \cdot B$, or (substituting for L) proportional to K_I^4. Since the probability of failure by cleavage is the complement of the joint probability of nonfailure of all the volume elements sampled by the crack-tip stress fields, the probability of failure must scale in proportion to the plastically deformed volume and, consequently, in proportion to K_I^4. Thus, the scatter in the cleavage fracture toughness of all ferritic steels is theoretically expected to be described by a Weibull distribution having a shape parameter of 4 (see Wallin 84), provided only that such a failure occurs under SSY conditions, which are characteristic of PTS loading of cracks in thick-walled RPVs (see Section 4.1.1). Figure 4-11 demonstrates that empirical evidence supports the theoretical expectation that the scatter in K_{Ic} and K_{Jc} data of all ferritic steels is well characterized by a Weibull distribution having a slope of 4.

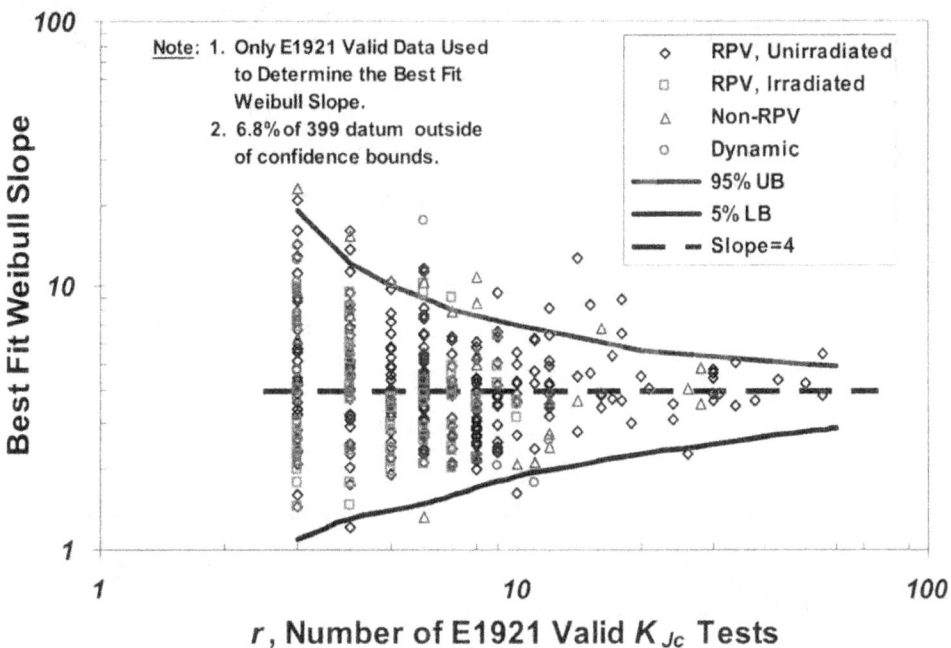

Figure 4-11 Comparison of Weibull shape parameters calculated from fracture toughness data with 5/95 percent confidence bounds on the expected shape parameter of 4 predicted by Wallin

4.2.1.2.2   Arrest Fracture Toughness

As outlined in Section 4.2.1, the occurrence or nonoccurrence of crack arrest depends upon the interaction of a rapidly evolving stress state in front of a running crack with the distribution of defects in the material that inhibits dislocation motion. Therefore, scatter in K_{Ia} data occurs as a consequence of the randomness in the distribution of barriers to dislocation motion throughout the material. Barriers to dislocation motion include vacancy clusters, interstitial clusters, coherent and semicoherent particles, and other dislocations. These dislocation-trapping defects are all of nanometer size and have interdefect spacings on the same size scale. The defects that control crack arrest are distributed at a much finer scale throughout the material than are the noncoherent particles responsible for crack initiation, which tend to have interdefect spacings of submicron order (1/10 micron). The possible variation in the local stress state over the microstructural distances that control crack arrest is therefore much smaller than that possible over the microstructural distances that control crack initiation. This smaller stress variation for crack arrest suggests that the scatter in K_{Ia} data should be smaller than in K_{Ic} data, a physically motivated expectation that agrees well with available empirical evidence, as shown in Figure 4-12 (see Kirk 02a). While this physical understanding is not yet sufficiently advanced to rationalize a distribution of crack arrest toughness values that is universal to all ferritic steels, available empirical evidence suggests that the distribution of crack arrest toughness values does not vary markedly among different ferritic steels (see Wallin 97).

4.2.2 Unirradiated Index Temperature and Transition Temperature Models

This section discusses models for unirradiated index temperatures and associated toughness transition models. Discussions of these two major parts of the crack initiation model are linked because index temperatures can only be discussed in the context of the characterization of transition fracture toughness from which they are derived. This discussion includes the RT_{NDT} model (see Section 4.2.2.1) as well as a best-estimate model enabled by the understandings detailed in Section 4.2.1 (see Section 4.2.2.2). Section 4.2.2.3 concludes with a description of how the NRC uses the best-estimate model together with RT_{NDT} to develop index temperature and toughness transition models for FAVOR, based on RT_{NDT}. Section 4.2.2.3 also includes a discussion of the classification and quantification of uncertainty associated with these FAVOR models.

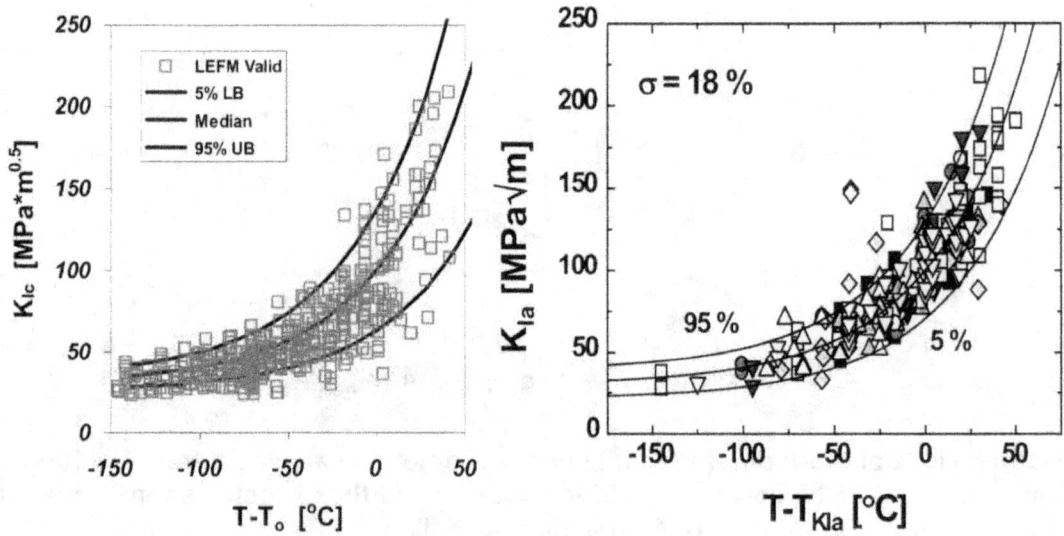

Figure 4-12 Comparison of scatter in crack initiation data (left) and in crack arrest data (right). Note that these figures are to the same scale, and that the median/mean curves have the same temperature dependence (i.e., $30 + 70 \cdot \exp\{0.019[T - T_{Norm}]\}$)

4.2.2.1 Current Model for Index Temperature (RT_{NDT})

RT_{NDT} is the index temperature used to position the American Society of Mechanical Engineers (ASME) K_{Ic} and K_{Ia} curves. However, as Section 4.2.2.1.1 describes, the development and use of RT_{NDT} can be traced back to studies of brittle fractures in ship steels, considerably predating its use by ASME to characterize nuclear RPV steels. Section 4.2.2.1.2 goes on to describe procedures for RT_{NDT} estimation that have been developed specifically for the characterization of nuclear RPV steels.

4.2.2.1.1   Historical Origins of RT_{NDT}

Section 4.2.1 demonstrated that the physical processes responsible for cleavage fracture initiation and arrest in ferritic steels make it possible to characterize the complete crack initiation and crack arrest transition fracture toughness behavior based only on an index temperature that locates the crack initiation transition curve on the temperature axis. While the physical understanding supporting the theoretical appropriateness of this approach has emerged only recently, index temperature approaches have been in use for over a century. The following subsections trace this development:

- Section 4.2.2.1.1.1 describes two laboratory scale test methods, the CVN test and the NDT test, that were developed to characterize transition index temperatures in ferritic steels.

- Section 4.2.2.1.1.2 describes the empirical correlations developed in the 1950s that relate the results of NDT tests to the fracture performance of structures.

- Section 4.2.2.1.1.3 describes how these relationships between NDT and structural performance were used in the 1960s in the development of toughness requirements for the ferritic steels employed in the construction of nuclear RPVs.

- Section 4.2.2.1.1.4 reviews ASME Code work from the early 1970s that relates CVN and NDT index temperatures to fracture toughness data.

4.2.2.1.1.1 Ductile to Brittle Transition and Associated Laboratory-Scale Test Methods. At the turn of the 20th century, it was already recognized that ferritic steels exhibit a fracture mode transition, from brittle fracture by transgranular cleavage at low temperatures to ductile fracture by microvoid coalescence at higher temperatures. In 1901, Charpy introduced the notched bar impact test (now codified as ASTM E23; see Charpy). Figure 4-13 shows the CVN specimen, which soon became a standard test for quantifying the ductile-to-brittle transition behavior of ferritic steels by performing these tests over a range of temperatures, as illustrated in Figure 4-14.

The spectacular failures of many of the liberty ships during World War II focused the technical community's attention on the importance of adequate toughness in structural steels and, in particular, on the importance of ensuring that ferritic steels operate at temperatures above their ductile-to-brittle transition temperature. The United States Naval Research Laboratory (NRL) conducted considerable research on the transition fracture performance of various test specimens, as well as of full-scale structures, from the end of World War II through the 1960s. One major outcome of this research was that the ductile-to-brittle transition temperature defined by the NDT test (now codified as ASTM E208) was a better indicator of structural performance than was the ductile-to-brittle transition temperature defined by CVN testing.

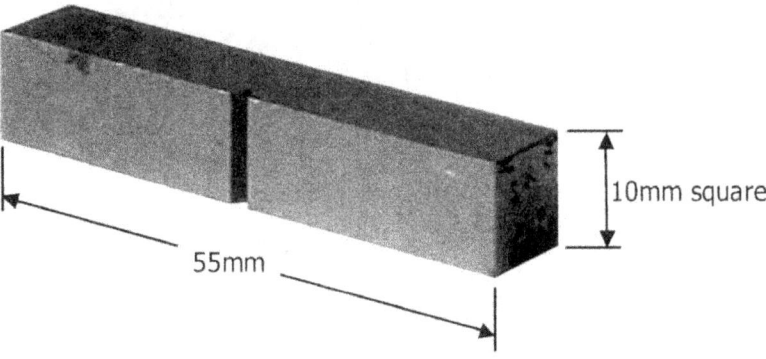

10mm square

55mm

Figure 4-13 Charpy V-notch impact test specimen

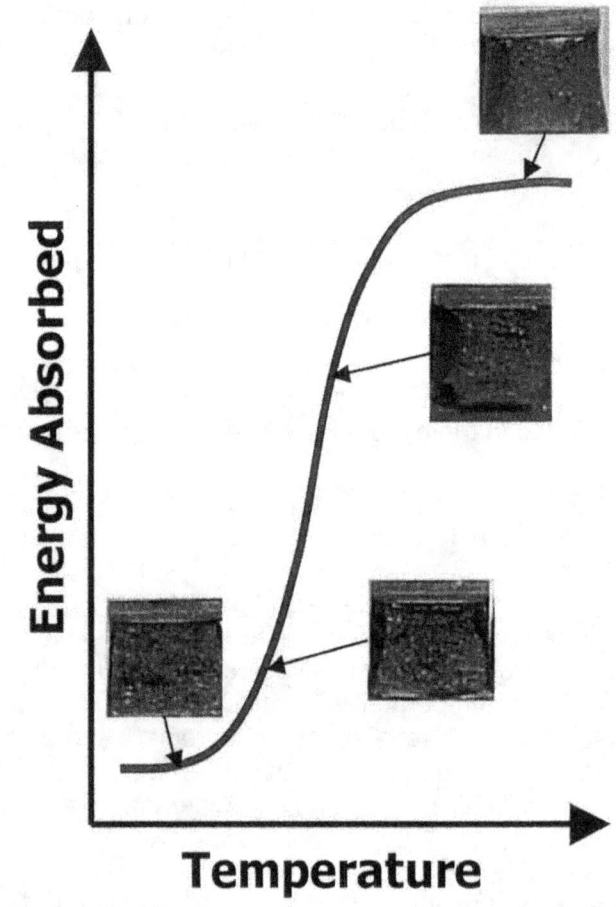

Figure 4-14 Charpy V-notch energy transition curve

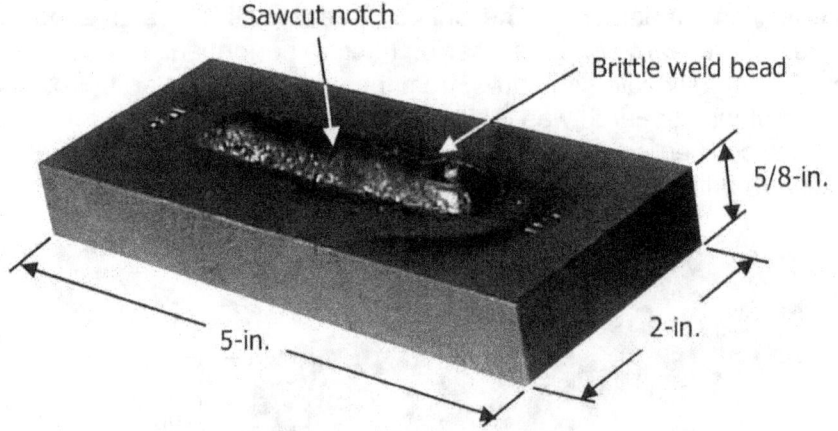

Figure 4-15 Nil-ductility temperature test specimen

The NDT test specimen, illustrated in Figure 4-15, features a notched brittle weld bead deposited on a 51 x 127 mm (2 x 5-in.) sample of steel that is 15.9 mm (5/8-in). thick. The NDT test involves impacting the unwelded side of the specimen with a falling weight and, as a consequence, bending the specimen by a fixed amount until the specimen hits a pair of mechanical stops. Performance of NDT tests over a range of temperatures (see Figure 4-16) defines the lowest temperature of no-break performance, which ASTM E208 defines as the NDT temperature (T_{NDT}).

4.2.2.1.1.2 Relationships between the NDT Temperature and Structural Performance. As mentioned in Section 4.2.2.1.1.1, the NRL team led by W.S. Pellini also performed extensive studies aimed at correlating T_{NDT} to the fracture performance of flawed steel structures, or of much larger specimens that were viewed as being representative of structures. The "failure analysis diagram" (FAD) illustrated in Figure 4-17 summarizes the result of these studies. The FAD, which is based on extensive quantities of fracture test data, service experience, and no small amount of "engineering judgment," illustrates the combined effects of flaw size, stress level, and temperature on the ability of ferritic steels to resist both crack initiation and crack arrest. For example, if one reads across the line at a stress level of 3/4-yield (labeled with an A-B-C-D on Figure 4-17), the diagram suggests that at temperatures below T_{NDT} (point "A" and to the left) an 203 mm (8-in.) long crack will initiate fracture. However, at temperatures above T_{NDT} progressively larger cracks are necessary to cause crack initiation because of the increase in fracture toughness that occurs with temperature. At T_{NDT} +15 °C (+27 °F) (point "B"), a 305 mm (12-in.)-long crack is necessary to initiate fracture at 3/4-yield; at T_{NDT} +22 °C (+40 °F) (point "C"), a 610 mm (24-in.)-long crack is necessary; and at temperatures above T_{NDT} + 26 °C (+47 °F) (point "D" and to the right), cracks, irrespective of their size, cannot initiate in cleavage at an applied nominal stress of 3/4-yield because this is the location of the crack arrest transition (CAT) curve. As illustrated in Figure 4-18, the CAT curve was established as a conservative upper bound to the temperatures required for crack arrest in wide steel plates held at constant temperatures and loaded to a constant remote stress. Section 4.2.2.1.1.3 discusses this CAT curve further because it provided the basis for the transition temperature and operational requirements initially established for commercial nuclear RPVs.

4.2.2.1.1.3 Use of the NDT CAT Curve in Establishing the Toughness Required of Nuclear RPV Steels. The early prototype and first generation commercial nuclear power plants were designed to ASME Code Sections I, "Rules for Construction of Boilers," or VIII, "Rules for Construction of Pressure Vessels," neither of which (at the time) placed toughness requirements on the steels used in vessel construction. Consequently, supplemental toughness requirements were developed. The first edition of Section III of the ASME Code, "Rules for Construction of Nuclear Facility Components," was published in 1963. This edition specified the lowest service temperature of the nuclear RPV as T_{NDT} +33 °C (+60 °F); Figure 4-17 reveals the origin of this requirement. At T_{NDT} +33 °C (+60 °F) the CAT curve passes through the yield stress level. Thus the ASME Section III requirement for a minimum operating temperature of T_{NDT} +33 °C (+60 °F) suggests that any flaws that remain in the vessel after fabrication cannot initiate (irrespective of their size) as long as the applied stresses remain below yield, as they were designed to do‡. On August 27, 1967, the predecessor of the NRC, the Atomic Energy Commission (AEC), issued a requirement that all beltline materials have a T_{NDT} of -12 °C (10 °F) or less. When combined with the ASME Section III requirement for a minimum operating temperature of T_{NDT} + 33 °C (+60 °F) and the empirical basis of Pellini's CAT curve, the August 27, 1967, AEC T_{NDT} requirement suggests that, at least at the beginning of life, a nuclear RPV could be fully pressurized at the ambient temperature without a risk of brittle failure.

---

‡ The 1963 ASME Section III toughness requirements considered neither the effects of accident loading nor the effects of irradiation embrittlement (other than implicitly through the use of generally conservative design principles).

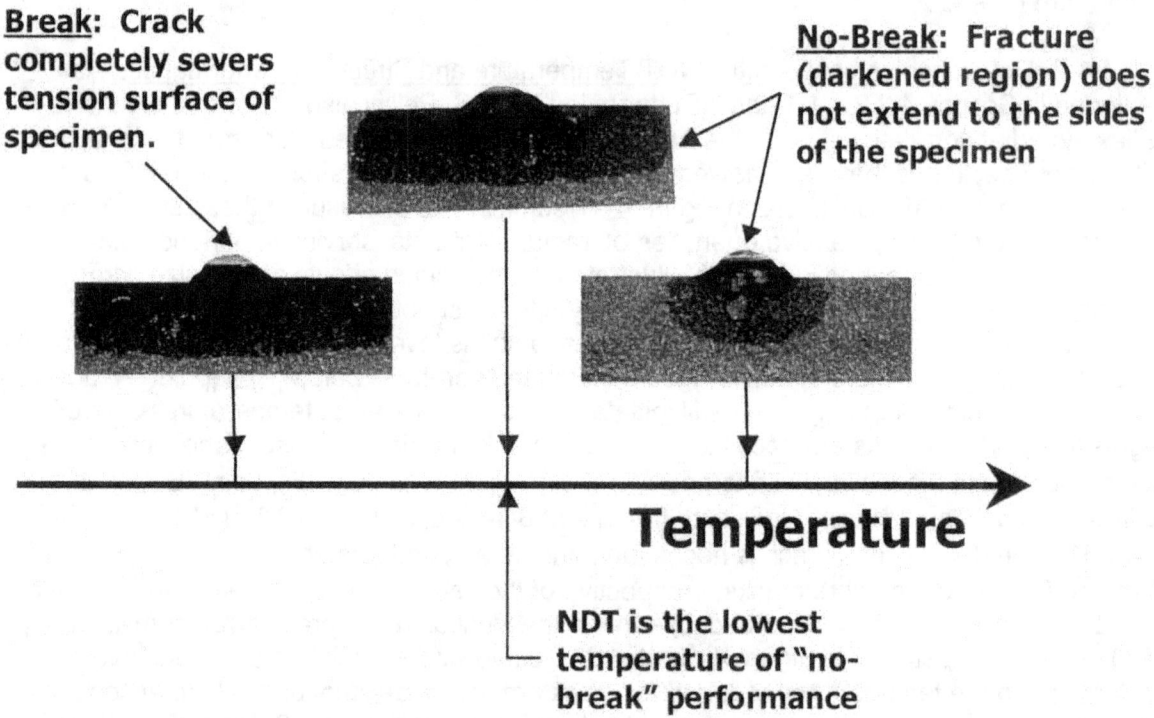

Break: Crack completely severs tension surface of specimen.

No-Break: Fracture (darkened region) does not extend to the sides of the specimen

Temperature

NDT is the lowest temperature of "no-break" performance

Figure 4-16 Definition of the NDT temperature

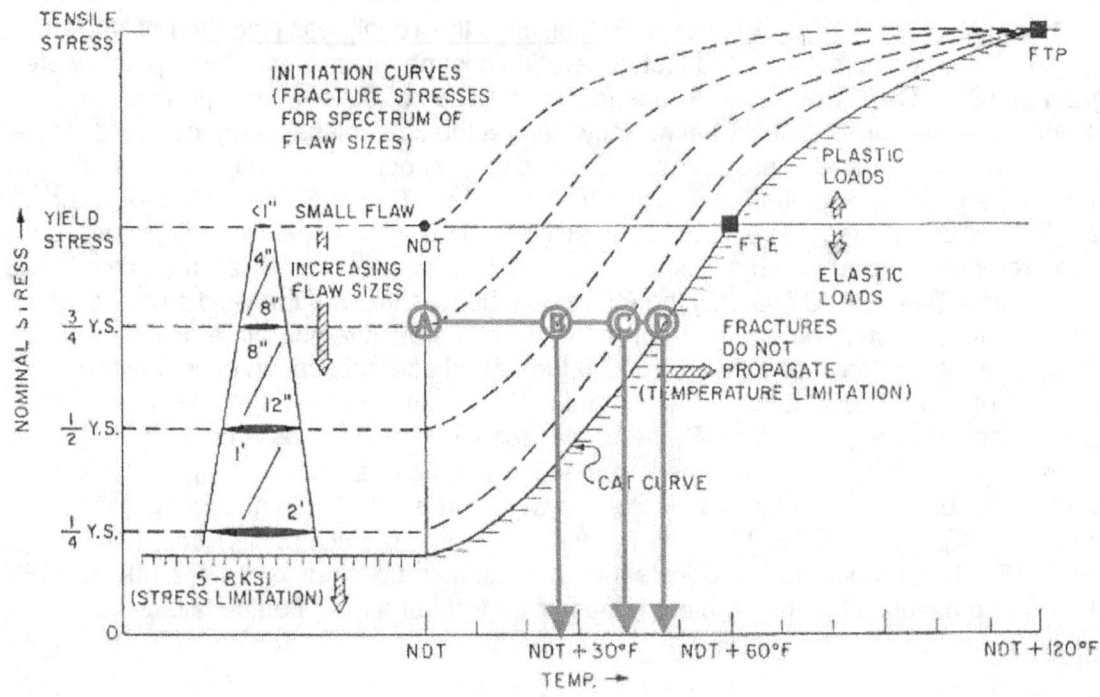

Figure 4-17 Generalized failure analysis diagram as presented by Pellini and Puzak (Pellini 63)

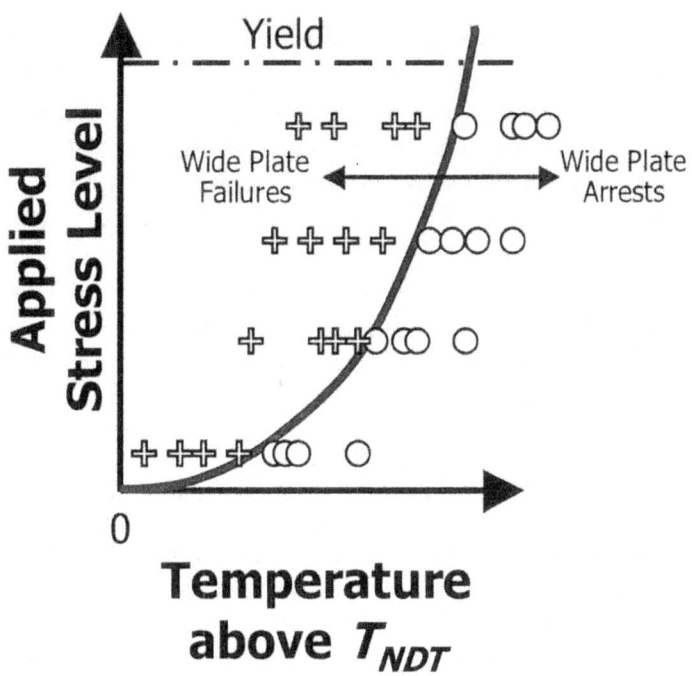

Figure 4-18 Construction of the CAT curve

4.2.2.1.1.4 <u>Definition of RT_{NDT} from CVN and NDT Data, and Its Use to Locate K_{Ic} and K_{Ia} Transition Curves</u>. In February 1971, the Pressure Vessel Research Council (PVRC) formed a task group to develop recommended toughness requirements for the ferritic materials in nuclear RPVs. In August of that same year, the task group delivered its recommendation to both ASME and the AEC. This recommendation was published in 1972 as Welding Research Council Bulletin 175 (see WRC175). The task group developed two curves as lower-bound representations of all of the LEFM-valid crack initiation toughness (K_{Ic}) and crack arrest toughness (K_{Ia}) data available at the time (these curves appear along with the data from which they were derived in Figure 4-19):

Eq. 4-1

$$K_{Ic} = 33.2 + 2.806 \cdot \exp[0.02 \cdot (T - RT_{NDT} + 100)]$$
$$K_{Ia} = 26.78 + 1.223 \cdot \exp[0.0145 \cdot (T - RT_{NDT} + 160)]$$

where

RT_{NDT} is (now) defined per ASME NB-2331 as $RT_{NDT} = MAX\{T_{NDT}, T_{35,50} - 60\}$,

T_{NDT} is the nil-ductility temperature determined by testing NDT specimens per ASTM E208, and

$T_{35,50}$ is the transition temperature at which Charpy-V notch (CVN) specimens tested per ASTM E23 exhibit at least 35 mills lateral expansion and 50 ft-lb absorbed energy[§].

Eq. 4-1 expresses stress intensity factor (K) values in units of ksi√in and temperature values in units of °F.

[§] The task group actually proposed a slightly different Charpy requirement, albeit one of similar intent. The Charpy requirement given is consistent with the current definition of RT_{NDT}.

4.2.2.1.2 Procedures for RT_{NDT} Estimation

Both Title 10, Section 50.61, of the *Code of Federal Regulations* (10 CFR 50.61) and Appendix G to 10 CFR Part 50 require an estimate of RT_{NDT} for each steel in the beltline region of the reactor. This section focuses attention only on the estimation of RT_{NDT} in the unirradiated condition, or $RT_{NDT(u)}$, because the effects of irradiation on index temperature shift appear in Section 4.2.3.

According to current regulations, there are three different methods of estimating $RT_{NDT(u)}$, as illustrated in Figure 4-20. The selection of which method to use is based on the information available. If both NDT and CVN data are available, the ASME NB-2331 method is used (see Sections 4.2.2.1.1.4 and 4.2.2.1.2.1). The two alternative methods are referred to as the MTEB 5.2 method and the generic method. The NRC developed these alternative methods during and shortly following the development of the technical basis for the current PTS rule to address situations in which sufficient CVN and NDT data were not available to estimate $RT_{NDT(u)}$ according to ASME NB-2331. As will become apparent in the following discussion, these different methods of estimating $RT_{NDT(u)}$ do not ensure the same degree of overestimation of the true fracture toughness transition temperature as is characteristic of ASME NB-2331 $RT_{NDT(u)}$ values.

The following sections summarize the preferred and the two alternative methods of $RT_{NDT(u)}$ estimation.

4.2.2.1.2.1 <u>ASME NB-2331—Preferred Method</u>. The RT_{NDT} is defined per ASME NB-2331 as follows:

Eq. 4-2 $$RT_{NDT} = MAX\{T_{NDT}, T_{35/50} - 60\}$$

where

T_{NDT} is the nil-ductility temperature determined by testing NDT specimens per ASTM E208 and

$T_{35,50}$ is the transition temperature at which CVN specimens tested per ASTM E23 exhibit at least 35 mills lateral expansion and 50 ft-lbs absorbed energy.

In WRC-175, the task group cited the following reasons for basing RT_{NDT} on both NDT and CVN data[**]:

(1)   The use of both NDT and CVN tests gives protection against the possibility of errors in conducting the tests or the reporting of test results.

(2)   The CVN requirements are expressed, in part, in terms of lateral expansion because this provides protection from variation in yield strength from initial heat treatment and the change in yield strength produced by irradiation.

[**] This list of four items is a direct quotation from WRC-175 with the exception that some nomenclature has been changed to ensure consistency with that used herein.

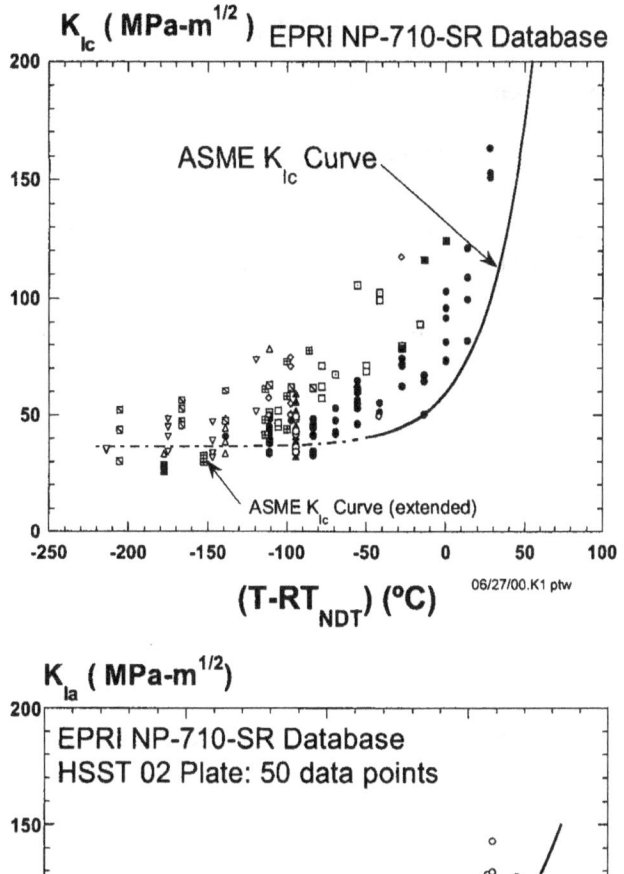

Figure 4-19 K_{Ic} and K_{Ia} toughness values used to establish the ASME K_{Ic} and K_{Ia} curves

(3)   The requirement of 0.9 mm (35 mills) lateral expansion and 68 J (50 ft-lbs) at T_{NDT} + 33 °C (60 °F), throughout the life of the component provides assurance of adequate fracture toughness at upper-shelf temperatures.

(4)   The CVN test at T_{NDT} + 33 °C (60 °F) serves to weed out nontypical materials such as those that might have low transition temperature but abnormally low energies on the upper shelf.

It can again be noted that the frequent use of the temperature T_{NDT} + 33 °C (60 °F) finds its origins in the Pellini CAT curve (see Figure 4-17).

4-21

4.2.2.1.2.2 MTEB 5.2—Alternative Method (for Plates and Forgings).

If only Charpy test data are available, the reference temperature value can be any one of a wide array of different index temperatures (e.g., T100, Tupper shelf, 16 °C (60 °F), T30, -18 °C (0 °F), T45, and T35/50 + 11 °C (20 °F)). As illustrated in Figure 4-20, the particular index temperature used depends on both the material type and the amount of data available. This myriad of index temperature measures for different materials and situations addresses all situations of limited data for plate and forging materials known to exist in 1981 (see NRC MTEB 5.2). Because MTEB 5.2 was developed as an alternative to the preferred ASME NB-2331 method, it is believed that all of these alternative index temperatures are more conservative than (i.e., higher than) an ASME NB-2331 RTNDT(u) value. Indeed, the authors of Enclosure A to SECY-82-465 characterize MTEB 5.2 RTNDT(u) values as being "not very satisfactory, because they are overconservative in some cases" (see SECY-82-465).

4.2.2.1.2.3 Generic—Alternative Method (for Welds). I

4.2.2.1.2.4

If no heat-specific material test data are available, a reference temperature value is determined based on the flux type of the weld material of interest. These "generic" RTNDT(u) values are the means of the populations of RTNDT(u) values measured according to ASME NB-2331 for different weld flux types. As defined in 10 CFR 50.61, generic RTNDT values can be either −49 °C (−56 °F) for welds made with Linde 1092, Linde 0124, and Linde 0091 fluxes, or -18 °C (0 °F) for welds made with Linde 80 flux††.

The non-Linde 80 value was established as the average of a data set of 92 $RT_{NDT(u)}$ measurements provided by Combustion Engineering for Linde 1092, Linde 0124, and Linde 0091 welds. More limited information available at the time for Linde 80 welds (25 $RT_{NDT(u)}$ measurements) was averaged to obtain the -18 °C (0 °F) value. Subsequently, the NRC staff approved a generic value of −21 °C (−5 °F) for Linde 80 welds.

4.2.2.2   Best-Estimate Model

Since the ASME Code K_{Ic} and K_{Ia} curves are located using RT_{NDT}, which is not determined from fracture toughness data but rather from CVN and NDT data, there is no assurance that K_{Ic} and K_{Ia} curves positioned using RT_{NDT} according to Eq. 4-1 will be located consistently with respect to actual fracture toughness (K_{Ic} and K_{Ia}) data for each and every heat of steel. Indeed, the definition of RT_{NDT} (see Eq. 4-2) suggests that consistent positioning relative to fracture toughness data is not an inherent characteristic of the RT_{NDT} model. To characterize and quantify the uncertainties in the initiation fracture toughness model (i.e., in both RT_{NDT} and K_{Ic}), it is necessary to define the current best estimate of the fracture toughness transition behavior and compare this estimate to the RT_{NDT} / K_{Ic} characterization. This section defines the current best-estimate model. It then compares this best estimate to the RT_{NDT}/K_{Ic} characterization in Section 4.2.2.3.

Earlier discussion of index temperature approaches to initiation fracture toughness characterization (see Section 4.2.1) described the current best physical understanding of the mechanisms responsible for cleavage fracture in the transition regime. This physical basis

---

†† Other generic $RT_{NDT(u)}$ values that have been established as a result of individual licensing actions; they are not reviewed in this report.

supports a temperature dependency of, and scatter in, initiation fracture toughness that is universal to all ferritic steels. Overwhelming empirical evidence testifies that these physical expectations manifest in reality (see Figure 4-8 and Figure 4-9). Currently, the Master Curve and its associated transition temperature, T_o, provide the best mathematical representation of these trends.

Wallin, working in collaboration with Sarrio and Törrönen, began to publish papers that would become the basis for what is now referred to as the "Master Curve" as part of his doctoral research work in 1984 (see Wallin, all citations). This work includes (1) a statistical model of cleavage fracture, and (2) a temperature dependency of fracture toughness common to all ferritic steels, in agreement with the physical expectations detailed in Section 4.2.1. Mathematically, these features are expressed as follows:

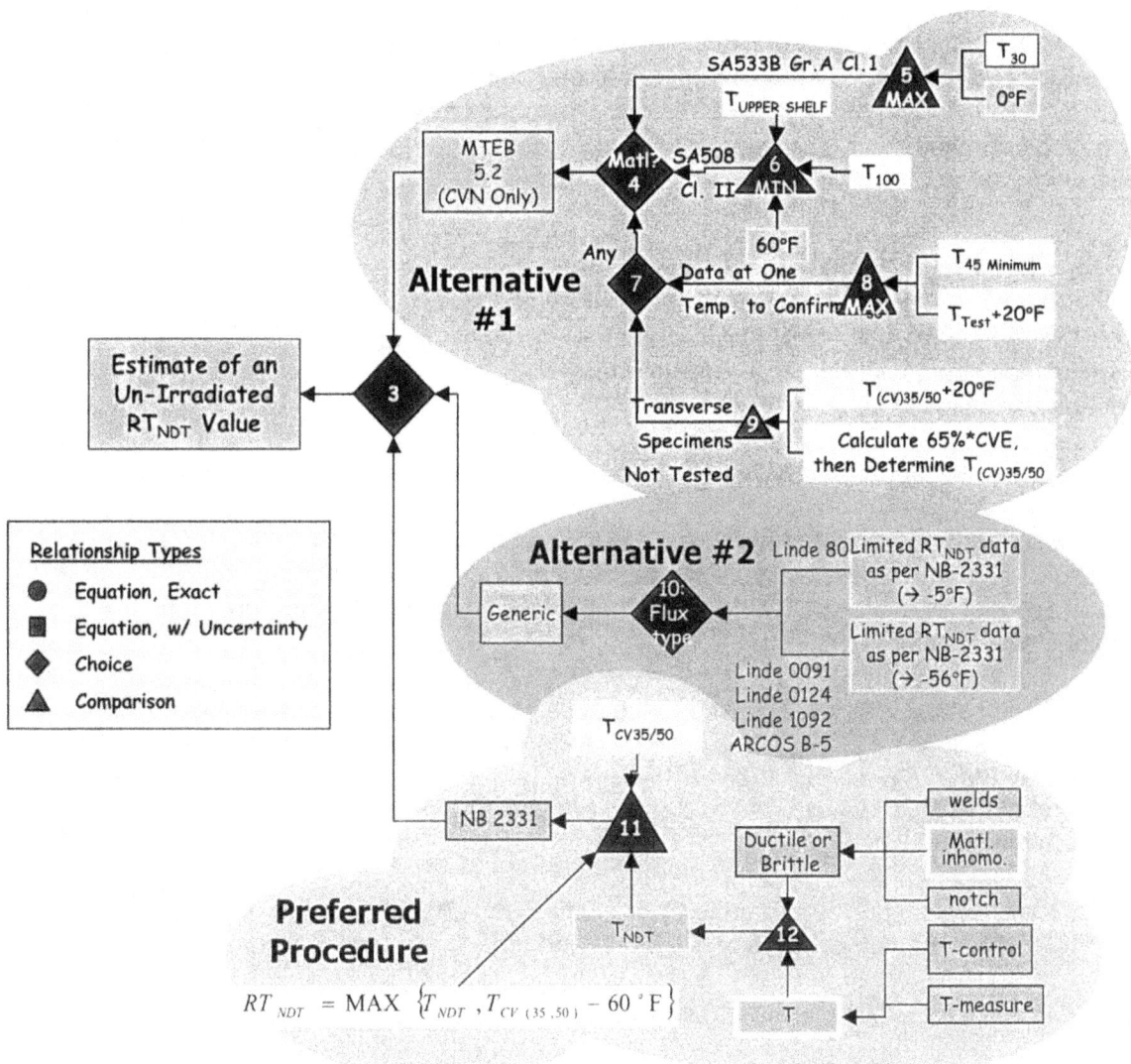

Figure 4-20 Diagram illustrating the different methods used currently to estimate a value of RT_{NDT} for an unirradiated RPV steel

4-23

Eq. 4-3 $\qquad K_{Jc(median)} = 27.30 + 63.71 \cdot \exp[0.0106(T - T_o)]$

Eq. 4-4 $\qquad P_f = 1 - \exp\left\{-\dfrac{B}{B_o}\left(\dfrac{K_I - 18.20}{K_o - 18.20}\right)^4\right\}$

Eq. 4-3 describes the temperature (T, in °F) dependency of the median fracture toughness ($K_{Jc(median)}$, in ksi√in). In this equation, temperature is normalized to the index temperature T_o, which is defined as the temperature at which the median toughness of a fracture specimen having the reference thickness (B_o, which is defined to be 25.4 mm (1-in.) is 100 MPa√m (91.01 ksi√in). Eq. 4-4 provides the three-parameter Weibull distribution that describes the distribution of toughness values about this median at all temperatures in transition. Of these three parameters, two are fixed—(1) the shape parameter is fixed at 4 and (2) the minimum value is set to 18.18 ksi√in. The parameter K_o, which corresponds to a 63.2 percent probability of failure, is determined by fracture toughness testing as described by American Society for Testing and Materials (ASTM) Standard E1921. K_o and $K_{Jc(median)}$ are related as follows:

Eq. 4-5 $\qquad K_{Jc(median)} = 0.9124(K_o - 18.20) + 18.20$

Figure 4-21 overlays Eq. 4-3 and Eq. 4-4 on cleavage fracture toughness data for a wide variety of ferritic steels, demonstrating the excellent agreement of the Wallin Master Curve model with experimental observations.

4.2.2.3 Model Used in FAVOR and Treatment of Uncertainties

Having described the physics underlying cleavage fractures (Section 4.2.1), the RT_{NDT} model (Section 4.2.2.1), and a best-estimate model (Section 4.2.2.2), the following sections will propose a model for use in FAVOR, and will both classify and quantify the uncertainties associated with this model.

Figure 4-22 illustrates that, as a consequence of its physical basis and its definition, T_o consistently positions a model (in this case the ASME K_{Ic} curve) relative to fracture toughness data in a way that RT_{NDT} cannot (because RT_{NDT} is not a measure of fracture toughness). Information of this type can be used to classify and quantify both the uncertainty in RT_{NDT} and the uncertainty in K_{Ic}.

The following sections discuss both the classification and quantification of uncertainty in the RT_{NDT}/K_{Ic}, first for RT_{NDT} (see Section 4.2.2.3.1) and then for K_{Ic} (see Section 4.2.2.3.2).

4.2.2.3.1 Index Temperature, RT_{NDT}

4.2.2.3.1.1 Uncertainty Classification. As discussed in Section 4.2.2.1.2, uncertainty arises in RT_{NDT} values for a number of reasons:

(1)   the conservative bias inherent to the ASME NB-2331 definition of RT_{NDT}

(2)   the myriad of methods (i.e., ASME NB-2331, generic, and MTEB 5.2) and transition temperatures (e.g., T_{NDT}, $T_{CV35/50}$, T_{30}, T_{45}, T_{100}) used to define RT_{NDT}

(3)   the limited data sets used to define generic RT_{NDT} values and to assess the appropriateness of the various transition temperatures used in RT_{NDT} definitions

(4)   the lack of prescription in the test methods (NDT and CVN) used to develop the properties that define RT_{NDT}, and the fact that neither the NDT nor the CVN test method actually measures a material property (making NDT and CVN data depend heavily on specimen geometry, preparation, and test method)

(5)   material variability

Of these sources of uncertainty, the first four clearly reflect uncertainty brought about by a limited state of knowledge, and so are epistemic in nature. Only uncertainty resulting from material variability can be regarded as irreducible (aleatory). However, this information alone is insufficient to determine if the uncertainty in RT_{NDT} is primarily aleatory or epistemic. To make this distinction, one must compare the RT_{NDT} index temperature to the index temperature associated with the best-estimate model for crack initiation toughness (T_o). As detailed in Section 4.2.2.2, this best-estimate model is supported by strong physical insights, which demonstrate that the trends the empirical data are expected, and more importantly are expected to apply to the range of material and irradiation conditions characteristic of PWR service. This, combined with the fact that T_o is estimated directly from fracture toughness data, and so, by definition, must relate to the same location on the transition temperature curve of every material, suggests that the epistemic uncertainty sources that plague RT_{NDT} do not influence T_o. Because the uncertainty in the best-estimate transition temperature is expected to be primarily aleatory, a comparison of T_o and RT_{NDT} values can be used to quantify the epistemic uncertainty associated with RT_{NDT}.

4.2.2.3.1.2 <u>Uncertainty Quantification</u>. Section 4.2.2.2 identified the Wallin Master Curve as the best-estimate model of initiation fracture toughness for ferritic steels. Because of the direct link between the Master Curve and the Master Curve transition temperature (T_o), T_o is the best estimate currently available of the transition fracture toughness of RPV steels, both before and after irradiation. Consequently, T_o is used to quantify the epistemic uncertainty associated with RT_{NDT}. The following sections first develop a Master Curve-based procedure for uncertainty quantification, and then modify this procedure to make it consistent with the constraint imposed on the PTS reevaluation effort that all models of fracture toughness and fracture toughness uncertainty be consistent with the principles of LEFM.

4.2.2.3.1.2.1 <u>Master Curve Procedure</u>. To account for the epistemic uncertainties in RT_{NDT}, one must quantify how far away from the measured fracture toughness data RT_{NDT} positions a model of fracture toughness, as illustrated in Figure 4-23. As detailed previously, the Master Curve transition temperature, T_o, best represents the true fracture toughness transition temperature. By definition, T_o represents the same point on the transition curve for every ferritic steel (i.e., the temperature at which the median fracture toughness of a 1-in.-thick specimen is 90.9 ksi√in). Thus, T_o must correspond to the position of fracture toughness data, rather than some model of the data located based on information other than fracture toughness, as is the case with RT_{NDT}-based models.

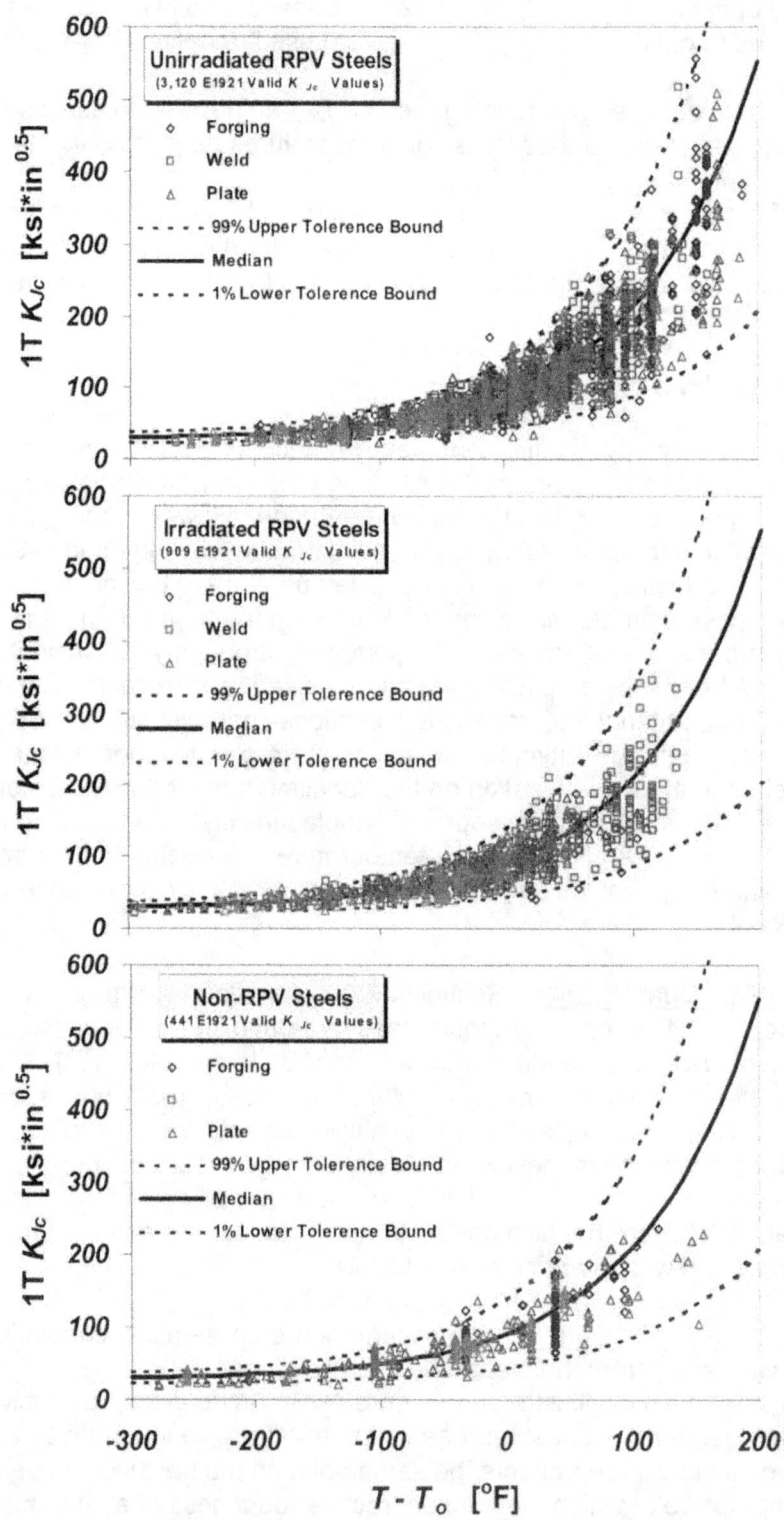

Figure 4-21 The uniform variation of cleavage fracture toughness with temperature noted by Wallin for (top) unirradiated RPV steels, (middle) irradiated RPV steels, and (bottom) other ferritic steels

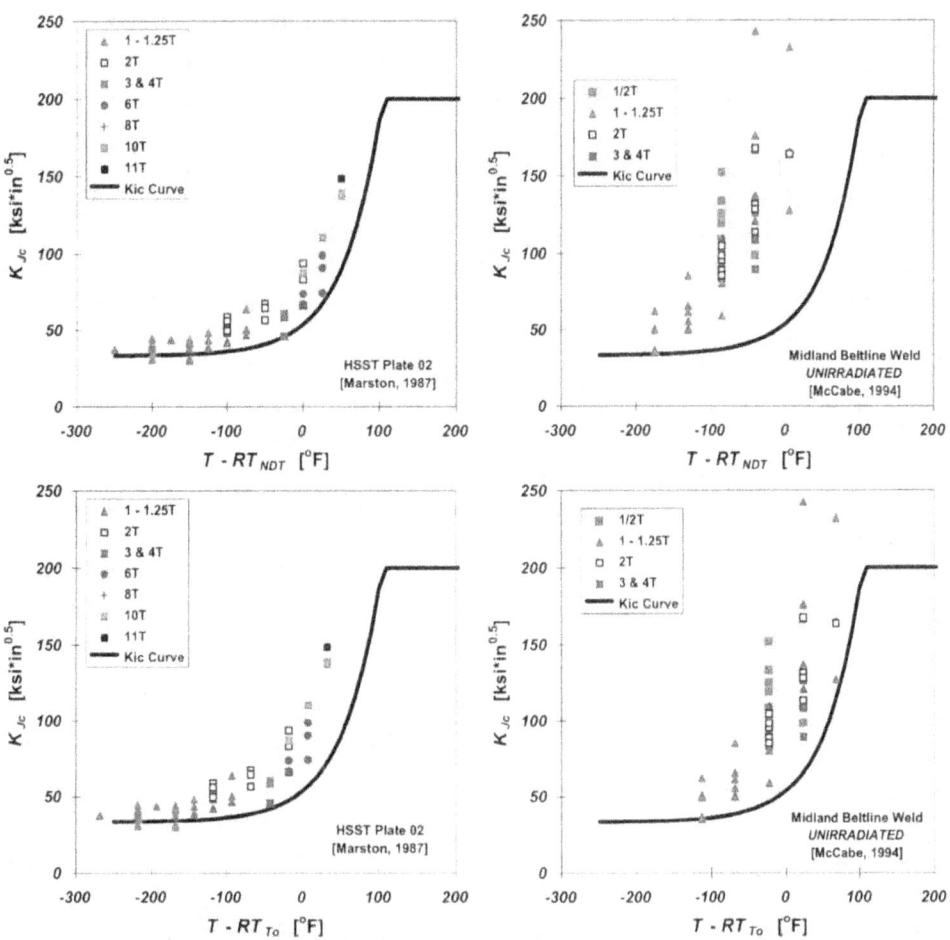

Figure 4-22 Comparison of how RT_{NDT} positions a transition toughness model (the K_{Ic} curve) relative to as-measured fracture toughness data (top) vs. how RT_{To} (defined from T_o based on ASME Code Case N-629) positions the same transition toughness model relative to as-measured fracture toughness data (bottom)

The NRC collected information from the literature where both RT_{NDT} and T_o were reported for the same RPV steels (see Table 4-1). Because T_o corresponds to the location of fracture toughness data by definition, this information allowed the NRC to quantify the uncertainty in RT_{NDT} as the simple difference between RT_{NDT} and T_o. Figure 4-24 shows the cumulative distribution function (CDF) constructed from these difference values ($\Delta RT = RT_{NDT} - T_o$), demonstrating that RT_{NDT} almost always provides a conservative estimate of the true fracture toughness transition temperature.

4.2.2.3.1.2.2 Modification of Master Curve Procedure for Consistency with LEFM.

While the Master Curve arguably reflects the best model of transition fracture toughness available today, its explicit treatment of size effects is inconsistent with the LEFM-based methods employed in FAVOR. Consequently, an alternative to the Master Curve-based procedure described in the preceding section is necessary. This alternative procedure avoids the explicit treatment of statistical size effects on cleavage fracture toughness adopted by the Master Curve model (see Eq. 4-4), thereby bringing it into compliance with the constraints imposed on toughness models used in the PTS reevaluation effort. This discussion focuses on the use of this procedure with the extended K_{Ic} fracture toughness database developed by Oak Ridge National Laboratory (ORNL) (see Bowman 00). The section concludes with a comparison of these two characterizations of epistemic uncertainty.

To avoid explicit treatment of size effects (Eq. 4-4), a procedure is developed that quantifies the bias, or epistemic uncertainty, in an RT_{NDT}-based fracture toughness model relative to K_{Ic} values that satisfy the ASTM E 399 plane-strain validity criteria. This procedure is applied to a database of 254 K_{Ic} values (see Bowman 00). Figure 4-25 depicts these data as a function of the normalized temperature T-RT_{NDT}, along with the ASME K_{Ic} curve and an adjusted lower-bounding K_{Ic} curve (see Nanstad 93). As illustrated in Figure 4-26, a temperature shift, ΔRT_{LB}, was determined for each of the 18 heats of RPV steel in this database by treating the adjusted lower-bounding K_{Ic} curve in Figure 4-25 as a 1-percent quantile curve and determining the temperature shift needed to make this curve coincident with the lowest K_{Ic} value for the heat of steel under consideration. This procedure is similar to the Master Curve-based procedure discussed in Section 4.2.2.3.1.2.1 insofar as measured fracture toughness data are regarded as the "truth" that the index temperature needs to represent, but it avoids an explicit treatment of size effects adopted by the Master Curve by relying on only LEFM-valid K_{Ic} values when defining quantifying epistemic uncertainty. Table 4-2 lists the ΔRT_{LB} value for each of the 18 heats of steel in the ORNL 99/27 database.

Figure 4-27 provides a CDF determined from these ΔRT_{LB} values, with the Weibull parameters of the CDF calculated using the Method of Moments point-estimators.

Eq. 4-6 $\Delta RT_{LB} = -40.02 + 124.88[-\ln(1-P)]^{0.51}$

This CDF quantifies the epistemic uncertainty in RT_{NDT} in a manner fully consistent with the constraints placed on toughness models used in the PTS reevaluation effort. Figure 4-27(c) also compares this quantification of epistemic uncertainty with that based on the Master Curve. This comparison illustrates that an implicit treatment of size effects produces an epistemic uncertainty quantification that lies between the Master Curve-based CDFs for 1T and 4T specimens, a placement that provides some sense of the average size of the fracture toughness specimens used in the definition of ΔRT_{LB}.

The adjustment to RT_{NDT} quantified by Figure 4-27 is based on the difference between RT_{NDT} values estimated using ASME NB-2331 procedures and LEFM-valid fracture toughness data. Consequently, in addition to the uncertainty associated with using RT_{NDT} to model the true fracture toughness transition temperature of the material (i.e., the difference between RT_{NDT} and the LEFM-valid data), the CDF in Figure 4-27 quantifies the combined uncertainty attributable to all of the sources at and to the right of node 11 on Figure 4-20, including the following:

- uncertainty in the ASME NB-2331 definition of RT_{NDT}
- uncertainty arising from material nonhomogeneity
- uncertainty in the CVN and NDT testing methodologies

Therefore, the CDF in Figure 4-27 represents the total epistemic uncertainty in RT_{NDT}. However, all RT_{NDT} values were not determined by the ASME NB-2331 method used to generate the data that support the ΔRT_{LB} adjustment factor. As discussed in Section 4.2.2.1.2 and illustrated in Figure 4-20, RT_{NDT} values are also determined using the MTEB 5.2 and generic methods. The use of the RT_{NDT} adjustment in Figure 4-27 to quantify the epistemic uncertainties in the MTEB 5.2 and generic RT_{NDT} values is appropriate for the following reasons:

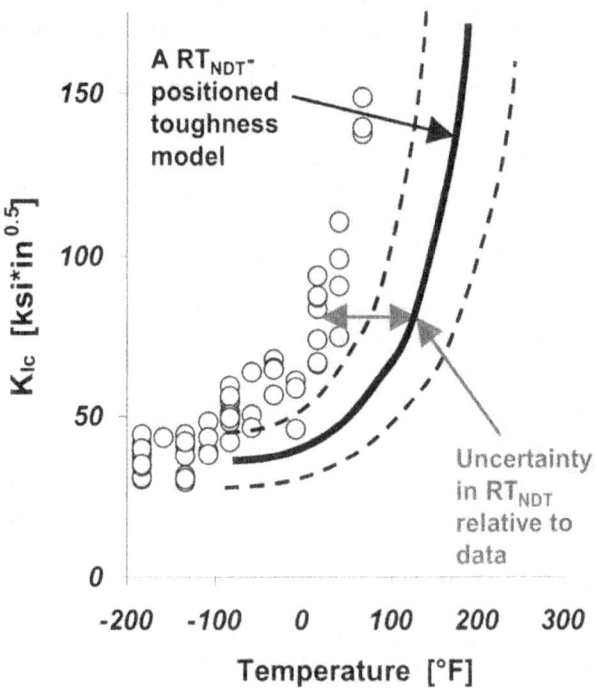

Figure 4-23 Illustration of how the error in an *RT$_{NDT}$*-based model of fracture toughness transition is determined

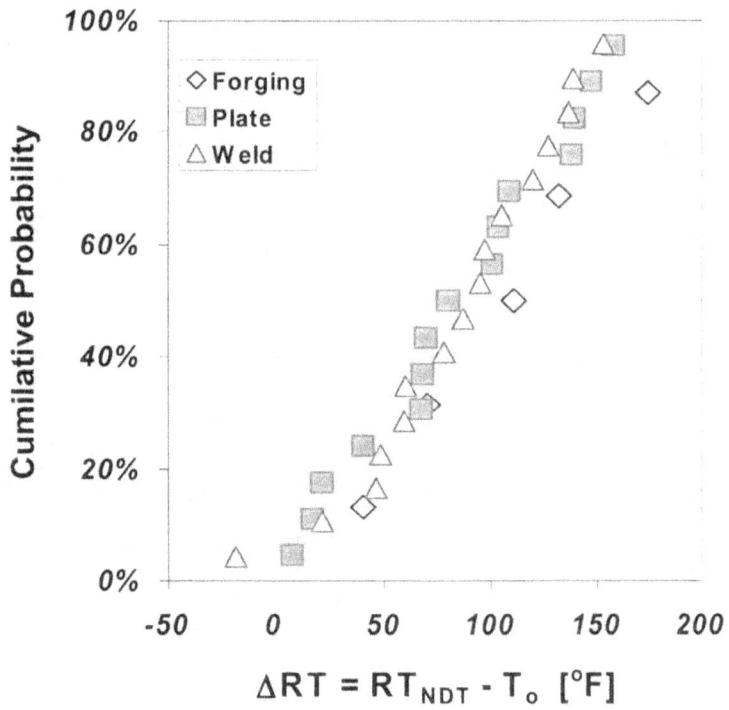

Figure 4-24 Cumulative distribution function showing the difference between T_o and RT_{NDT}

Table 4-1 Summary of Unirradiated RPV Materials Having Both RT_{NDT} and T_o Values Available

Author	Year	Product Form	Spec	Material Designation	T_o [°F]	RT_{NDT} [°F]	$RT_{NDT} - T_o$ [°F]
Iwadate, T.	1983		A508 Cl. 3		-54	-13	41
Marston, T.U.	1978		A508 Cl. 2		-6	65	71
Marston, T.U.	1978	Forging	A508 Cl. 2		-60	51	111
VanDerSluys, W.A.	1994		A508 Cl. 3		-154	-22	132
Marston, T.U.	1978		A508 Cl. 2		-124	50	174
McGowan, J.J.	1988		A533B Cl. 1	HSST 02	-8	0	8
Marston, T.U.	1978		A533B Cl. 1	HSST 02	-17	0	17
Marston, T.U.	1978		A533B Cl. 1	HSST 01	-2	20	22
Ahlf, Jurgen	1989		A533B Cl. 1	HSST 03	-21	20	41
Onizawa, Kunio	1999		A533B Cl. 1		-99	-31	68
Ishino, S.	1988		Generic Plate		-81	-13	68
CEOG	1998		A533B Cl. 1		-85	-15	70
Link, Richard	1997	Plate	A533B Cl. 1	HSST 14A	-70	10	80
McCabe, D.E.	1992		A533B Cl. 1	HSST 13A	-110	-9.4	100
Onizawa, Kunio	1999		A533B Cl. 1		-152	-49	103
Ishino, S.	1988		Generic Plate		-131	-22	109
CEOG	1998		A533B Cl. 1		-133	5	138
Marston, T.U.	1978		A533B Cl. 1		-74	65	139
Morland, E	1990		A533B Cl. 1		-142	5	147
Ingham, T.	1989		A533B Cl. 1		-154	5	159
Ishino, S.	1988				-39	-58	-19
Ishino, S.	1988				-98	-76	22
CEOG	1998				-126	-80	46
Ramstad, R.K.	1992			HSST 73W	-78	-29.2	48
McCabe, D.E.	1994			Midland Nozzle	-32	27	59
Ramstad, R.K.	1992			HSST 72W	-70	-9.4	60
CEOG	1998				-138	-60	78
CEOG	1998	Weld			-136	-50	86
Williams.	1998			Kewaunee 1P3571	-144	-50	94
McCabe, D.E.	1994			Midland Beltline	-70	27	97
Marston, T.U.	1978				-105	0	105
CEOG	1998				-139	-20	119
CEOG	1998				-157	-30	127
CEOG	1998				-186	-50	136
CEOG	1998				-189	-50	139
Williams, J.	1998				-203	-50	153

Table 4-2 Three Reference Transition Temperatures Defined Using the ORNL 99/27 K_{Ic} Database

Property Set ID	Material Description	Product Form	Sample Size	Reference Temperatures			Uncert. Terms	
				$RT_{NDT(u)}$[1]	T_0[2]	RT_{LB}[3]	$RT_{NDT(u)} - T_0$	ΔRT_{LB}
			N	(°F)	(°F)	(°F)	(°F)	(°F)
1	HSST 01	weld	8	0	-105	-64.3	105	64.3
2	A533 Cl. 1	weld	8	0	-57	10.9	57	-10.9
3	HSST 01	plate	17	20	-1	-77.8	21	97.8
4	HSST 03	plate	9	20	31	-71.5	-11	91.5
5	A533 Cl. 1	plate	13	65	-74	-121.4	139	186.4
6	HSST 02	plate	69	0	-17	-2.1	17	2.1
7	A533B	weld	10	-45	-151	-187.2	106	142.2
8	A533B	weld/HAZ	6	0	-132	-162.4	132	162.4
9	A508 Cl. 2	forging	12	50	-124	-97.6	174	147.6
10	A508 Cl. 2	forging	9	51	-60	0.9	111	50.1
11	A508 Cl. 2	forging	10	65	-55	10.4	120	54.6
12	HSSI 72W	weld	12	-9.4	-70	-15.4	60.6	6
13	HSSI 73W	weld	10	-29.2	-78	-67.6	48.8	38.4
14	HSST 13A	plate	43	-9.4	-109	-42.6	99.6	33.2
15	A508 Cl. 3	forging	6	-13	-46	-11.3	33	-1.7
16	Midland Nozzle	weld	6	52	NA	-37.4	NA	89.4
17	Midland Beltline	weld	2	23	NA	-58.9	NA	81.9
18	Plate 02 4th Irr.	plate	4	0	NA	-62.3	NA	62.3

[1] Bowman, K.O. and P.T. Williams, "Technical Basis for Statistical Models of Extended K_{Ic} and K_{Ia} Fracture Toughness Databases for RPV Steels," ORNL/NRC/LTR-99/27, Oak Ridge National Laboratory, February 2000

[2] Kirk, M., et al., "Bias and Precision of T_0 Values Determined Using ASTM Standard E 1921-97 for Nuclear Reactor Pressure Vessel Steels," *Effects of Radiation on Materials: 19th International Symposium, ASTM STP 1366*, M.L. Hamilton, et al., eds., American Society for Testing and Materials, West Conshohocken, PA, pp. 143–161, 2000.

[3] Unpublished calculations by J.G. Merkle, August 7, 2000.

Table 4-3 Summary of Generic RT_{NDT} Values in the RVID Database

Material Group	Generic RT_{NDT} Value (mean) (°F)	Standard Deviation (°F)
CE Welds	-56	17
B&W Welds	-5	17
B&W Plates	1	26.9
B&W Forgings	3	31
Weld WF-25	-7	20.6

- The MTEB 5.2 method was developed to produce RT_{NDT} values conservative to (i.e., higher than) the ASME NB-2331 RT_{NDT} values. Indeed, the authors of Enclosure A to SECY-82-465 characterize MTEB 5.2 $RT_{NDT(u)}$ values as "not very satisfactory, because they are overconservative in some cases" (see SECY-82-465). Thus, use of the CDF in Figure 4-27 will underestimate the epistemic uncertainties in a MTEB 5.2 RT_{NDT} value (a conservative treatment). Sufficient evidence does not currently exist to establish a procedure that can be used to account for this additional conservatism.

- Generic RT_{NDT} values in the Reactor Vessel Integrity Database (RVID) represent averages of ASME NB-2331 RT_{NDT} values. However, the CDF of Figure 4-27 was developed based on bounding (i.e., ASME NB-2331) RT_{NDT} values, so its direct use to correct for the epistemic uncertainties associated with the generic RT_{NDT} values found in RVID is not appropriate. In this situation, a step is necessary to recover from the generic RT_{NDT} value one of the bounding values from which it was derived before using Eq. 4-6 to estimate the epistemic uncertainty in this RT_{NDT} value. Because the generic RT_{NDT} values represent averages of ASME NB-2331 RT_{NDT} values (i.e., bounding values), estimating a bounding RT_{NDT} value from a generic one is a simple matter of randomly selecting an RT_{NDT} value from a normal distribution having the same mean and standard deviation as the data set originally used to establish the generic RT_{NDT} value. Table 4-3 summarizes the generic RT_{NDT} values currently found in the RVID database.

4.2.2.3.2   Crack Initiation Toughness

4.2.2.3.2.1 Uncertainty Classification. The distribution of noncoherent particles throughout the BCC iron lattice alone establishes the scatter in K_{Ic} data (see Natishan 99a). It is possible, at least in principle, to know if a noncoherent particle exists at a particular point in the matrix. This might suggest an epistemic nature to K_{Ic} scatter, were it not for the fact that K_{Ic} does not exist as a point property. A K_{Ic} value always has a size scale associated with it, which is the plastically deformed volume. Upon loading, the presence of the crack elevates the stress state along the entire length of the crack front to the point that dislocations begin to move in the surrounding volume of material, which contains a distribution of barriers to their motion (e.g., noncoherent particles, grain boundaries, and twin boundaries). Sufficient accumulation of dislocations at a barrier can elevate the local stress state sufficiently to initiate a crack in the barrier, and, if the criteria for fracture are satisfied, propagate the crack through the entire surrounding test specimen or structure. Thus, the existence of a particular dislocation barrier at a particular location does not control K_{Ic}. Rather, K_{Ic} is controlled by the distribution of these barriers throughout the lattice, and how this distribution interacts with the distribution of elevated stresses along the crack front. Because the distribution of these barriers throughout the lattice is random and occurs at a size scale below that considered by the K_{Ic} model of toughness, the uncertainty in K_{Ic} data is irreducible. For this reason, FAVOR models the uncertainty in K_{Ic} as aleatory.

4.2.2.3.2.2 Uncertainty Quantification. Use of RT_{NDT} as a temperature-indexing parameter introduces epistemic uncertainty into models of K_{Ic} scatter when these models are derived empirically from experimental databases that include multiple heats of steel having different RT_{NDT} values. Thus, a purely empirical derivation of K_{Ic} models and uncertainty measures produces a mixture of aleatory and epistemic uncertainties. Proper mathematical treatment of mixed uncertainties is not currently possible using PRA techniques, suggesting the need for a different approach to uncertainty quantification.

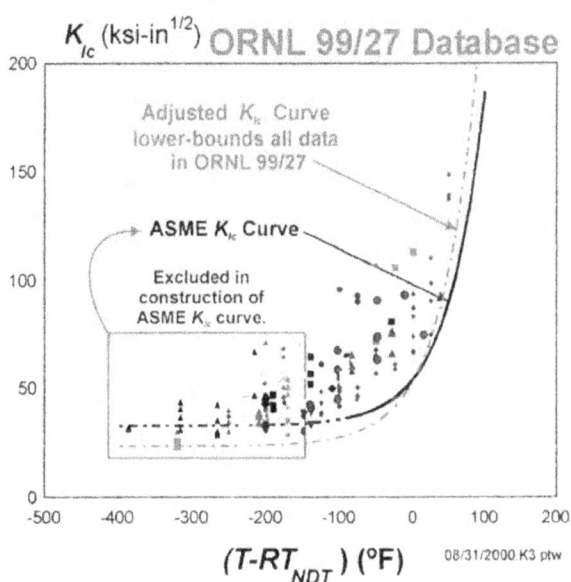

Figure 4-25 Extended K_{Ic} fracture toughness database (ORNL/NRC/LTR-99/27) (Bowman 00) of ASTM E399 valid data compared with adjusted ASME K_{Ic} curve (Nanstad 93)

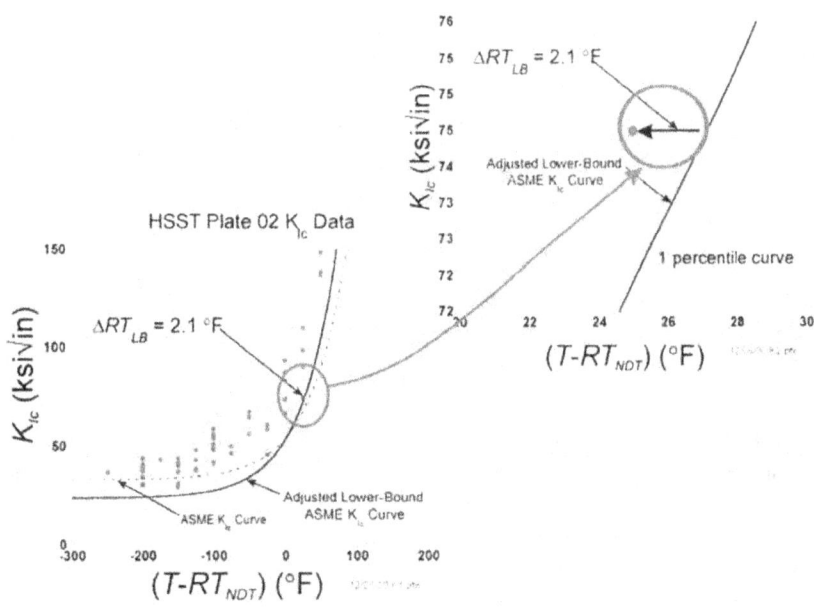

Figure 4-26 Illustration of the lower bounding methodology used to generate the uncertainty term (ΔRT_{LB}) for $RT_{NDT(u)}$

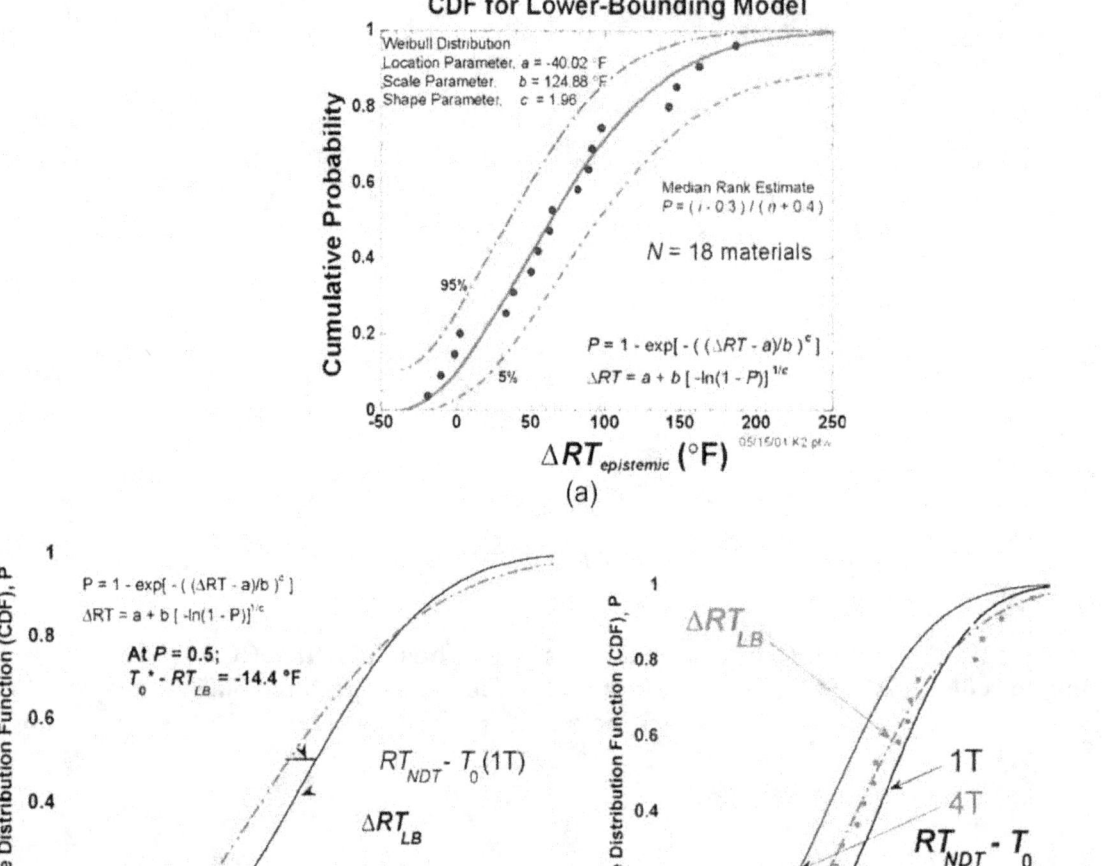

Figure 4-27 (a) Illustration of ΔRT_{LB} that quantifies both the epistemic uncertainty in $RT_{NDT(u)}$ and the intentional bias in $RT_{NDT(u)}$ values. (b and c) Comparison of ΔRT_{LB} adjustment with Master Curve-based ($RT_{NDT}\text{-}T_o$) adjustment.

As summarized in Section 4.2.2.2, a physical understanding of cleavage fractures demonstrates that the uncertainty (scatter) in K_{Ic} data is expected to follow a Weibull distribution, having a shape parameter of 4 and some finite lower bound value. This distribution was therefore assumed when fitting a data set of 254 LEFM-valid K_{Ic} values from 18 heats of RPV steel to establish the temperature dependence of K_{Ic} relative to the normalized temperature $T\text{-}RT_{LB}$ (see Figure 4-26 for RT_{LB} definition and Table 4-2 for RT_{LB} values). Figure 4-28 illustrates this best-fit model, which describes the aleatory uncertainty in K_{Ic}. Mathematically, this K_{Ic} model is as follows:

Eq. 4-7
$$K_{Ic}(\Delta T) = a_{K_{Ic}}(\Delta T) + b_{K_{Ic}}(\Delta T)\left[-\ln(1-P)\right]^{1/c_{K_{Ic}}} \quad \text{for } 0 \le P < 1$$

where

K_{Ic} is in ksi$\sqrt{\text{in}}$,

4-34

ΔT is $(T\text{-}RT_{LB})$, in °F, and
P is the fracture probability.

$$a_{K_{lc}}(\Delta T) = 19.35 + 8.335 \exp\left[0.02254(\Delta T)\right] \ [\text{ksi}\sqrt{\text{in.}}]$$

$$b_{K_{lc}}(\Delta T) = 15.61 + 50.132 \exp\left[0.008(\Delta T)\right] \ [\text{ksi}\sqrt{\text{in.}}]$$

$$c_{K_{lc}} = 4$$

$$-250°F \pounds \ \Delta T \ \pounds +50°F$$

4.2.2.4   Summary of Model and Uncertainty Treatment

Figure 4-29 combines the diagrammatic representation of the unirradiated index temperature and the crack initiation toughness transition models taken from Figure 4-6 with the specific models and input values recommended for use in the preceding sections. Uncertainties enter the model in the following places:

- When a generic (mean) value of $RT_{NDT(u)}$ is used, the epistemic uncertainty in this value is simulated by drawing a value from a standard normal distribution having the standard deviations given in Table 4-3. This sampled value is added to or subtracted from the generic $RT_{NDT(u)}$ value (also given in Table 4-3) to obtain the $RT_{NDT(u)}$ value used by FAVOR for a particular simulation run.
- The intentional conservative bias in $RT_{NDT(u)}$ is accounted for by sampling from the cumulative distribution function given in Figure 4-27(a). This sampled value ($\Delta RT_{epistemic}$) is then subtracted from $RT_{NDT(u)}$ to obtain a best-estimate value of the unirradiated index temperature for a particular simulation run.
- Figure 4-28 depicts the aleatory uncertainty in K_{lc} and its variation with temperature. This K_{lc} distribution is propagated throughout the rest of the model during each simulation run.

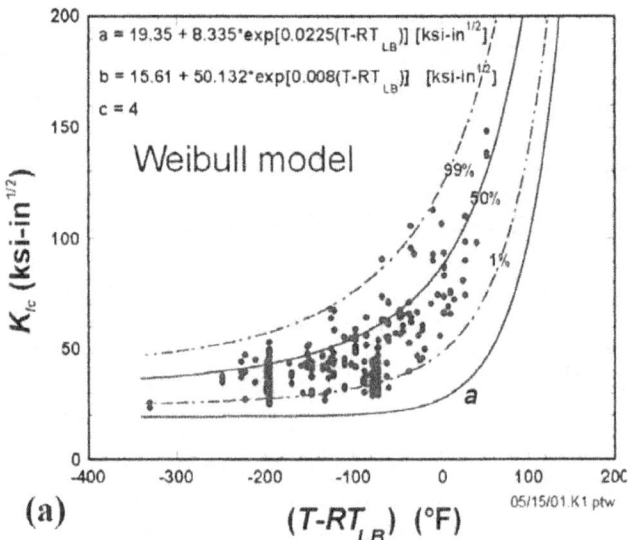

(a)

Figure 4-28 K_{lc} **model proposed for use in the PTS reevaluation effort. In fitting the model to the data, the Weibull shape parameter (c) was fixed at 4 while the minimum and median values (a and b, respectively) were defined based on the data.**

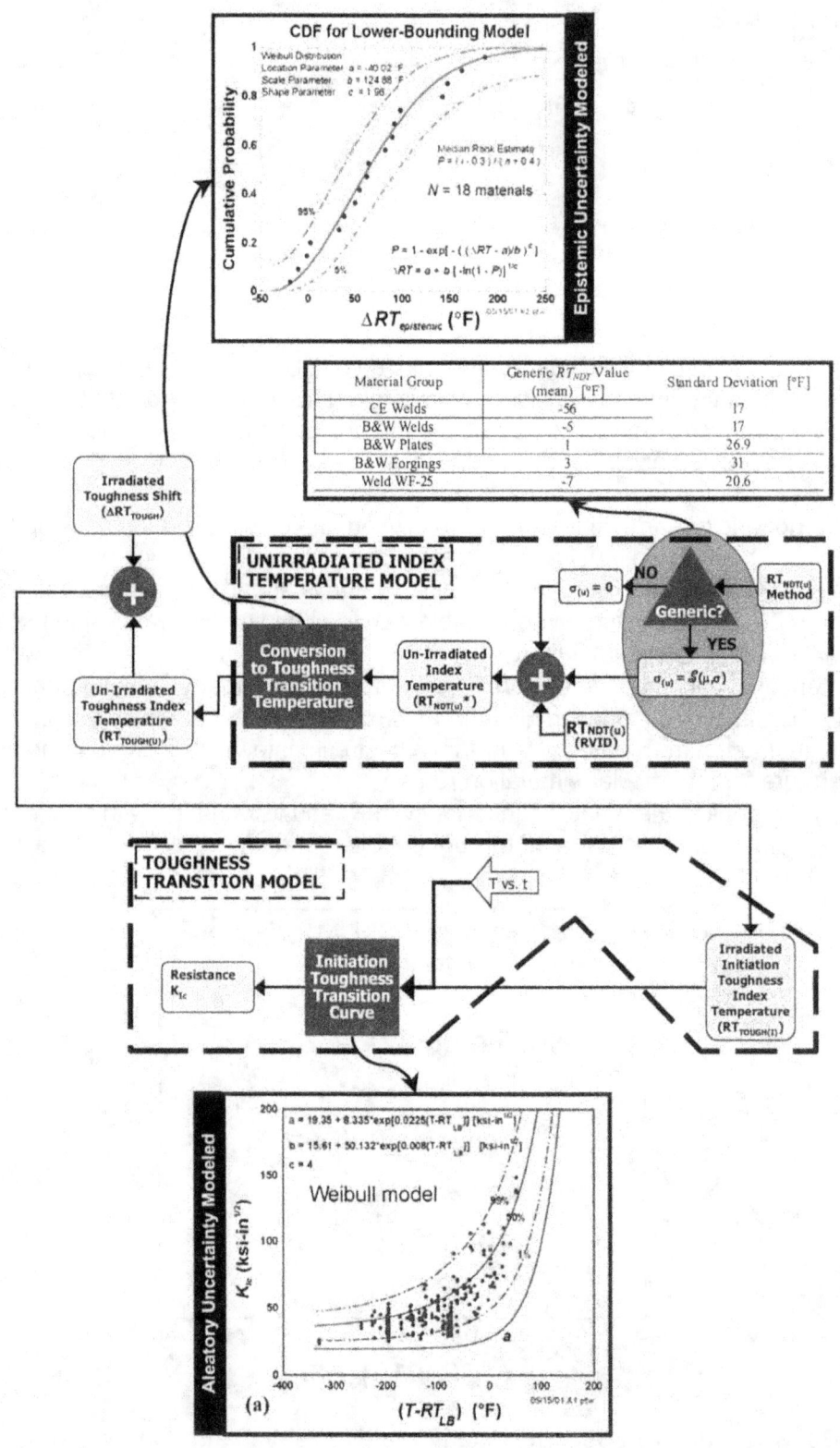

Figure 4-29 Diagrammatic summary of the FAVOR crack initiation model showing the recommended models, input values, and uncertainty treatment. Nongeneric values of $RT_{NDT(u)}$ can be found in the RVID database; these values are also provided in Appendix C to this document for the PWRs studied in the PTS reevaluation effort.

4.2.3 Index Temperature Shift Model

This section describes the model that FAVOR uses to estimate the shift in the toughness index temperature produced by irradiation. This model, illustrated in Figure 4-30, includes the following three submodels:

(1)   A fluence and attenuation model estimates the degree to which the fluence on the inner diameter of the RPV reduces through the RPV wall. Inner diameter fluences are predicted using the procedures of Regulatory Guide (RG) 1.190, "Calculational and Dosimetry Methods for Determining Pressure Vessel Neutron Fluence," which are described in detail elsewhere (see RG 1.190).

(2)   Another model estimates the degree to which the Charpy V-notch 30 ft-lb transition temperature (ΔT_{30}) increases because of irradiation damage, and the effect of both compositional and exposure variables on the degree of shift expected. This step (estimating the Charpy shift as a precursor to estimating the toughness shift) is necessary because at the current time insufficient experimental information exists from power reactor surveillance programs to support estimation of toughness shift directly from compositional and exposure variables.

(3)   A third model converts the shift in Charpy transition temperature to an estimate of the shift in toughness transition temperature necessary to position the fracture toughness transition curve of the irradiated material.

The following sections discuss the treatment of fluence uncertainty and the attenuation model. They also include information on the physical mechanisms responsible for irradiation damage of ferritic RPV steels that is needed to support the discussions of both the Charpy shift model and the toughness shift conversion model that follows. This section concludes with a summary of the model used in FAVOR and the uncertainty treatment associated with this model.

4.2.3.1   Fluence and Attenuation Model

4.2.3.1.1   Fluence Model

As illustrated in Figure 4-30, the fluence and attenuation model begins with an estimate of the azimuthal and axial variation of fluence on the inner diameter of the RPV (see Figure 4.31 and Figure 4.32 for example results). Neutron fluence transport calculations based on RG 1.190 provide these estimates. The calculations were performed in (r, θ) and (r, z) geometry using the DORT (see DORT) discrete ordinates transport code and the BUGLE-93 (see BUGLE) 47-neutron group ENDF/B-VI nuclear cross sections and fission spectra. The calculations employed an S_8 angular quadrature set, and a P_3 Legendre expansion represented the scattering cross sections.

The calculational models extended radially from the core out to the primary (concrete) biological shield and over an axial height from 1 foot below to 1 foot above the active fuel. The model retained the octant core symmetry and represented a 45° azimuthal sector of the geometry. The calculations included a detailed representation of the core/internals/vessel materials and geometry based on plant-specific information provided by the licensee and/or fuel vendor. The models incorporated the dimensions and region-specific material compositions of the core, barrel, thermal shield, vessel, and biological shield. The downcomer water density was determined using the inlet and outlet coolant temperatures and system pressure.

The calculations included a detailed modeling of the core neutron source using the MESH code (see MESH), which accounted for the reduced power in the fuel pins close to the core reflector, as well as the increased number and higher energy of the neutrons produced in the high burnup fuel assemblies. The source was based on the plant-specific operating data (i.e., the core thermal power operating history; the fuel assembly-wise power, burnup, and axial power distribution; and the peripheral fuel assembly pin-wise power distribution).

4.2.3.1.2 Fluence Uncertainty

The uncertainty in fluence values estimated by RG 1.190 calculations arises from the following sources:

- uncertainty in the vessel diameter
- uncertainty in the peripheral neutron source
- uncertainty in the core inlet temperature
- uncertainty in the neutron cross sections
- uncertainty in the computational method itself

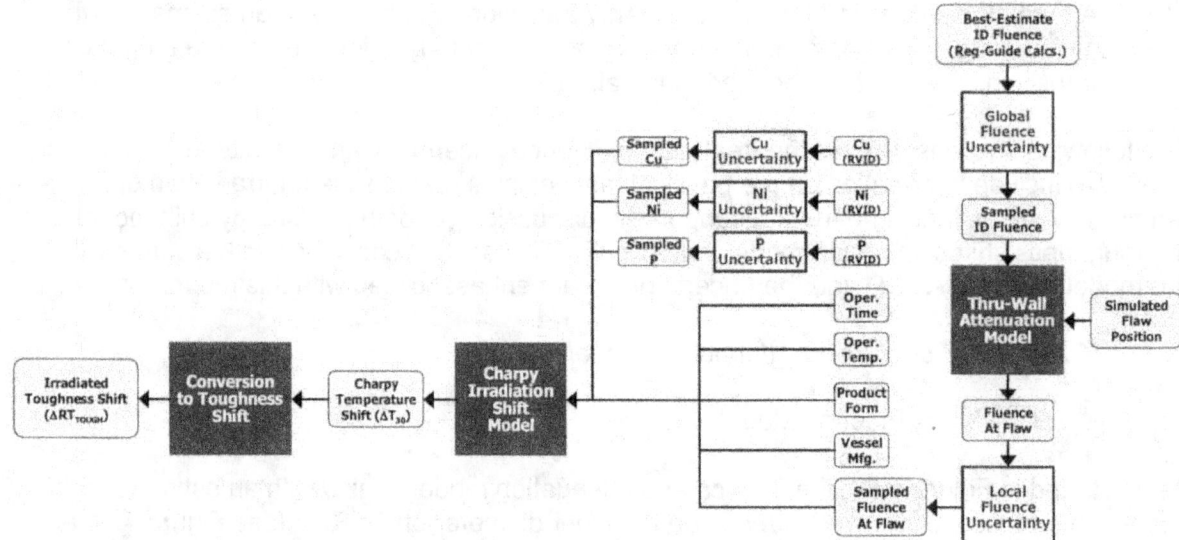

Figure 4-30 Index temperature shift model

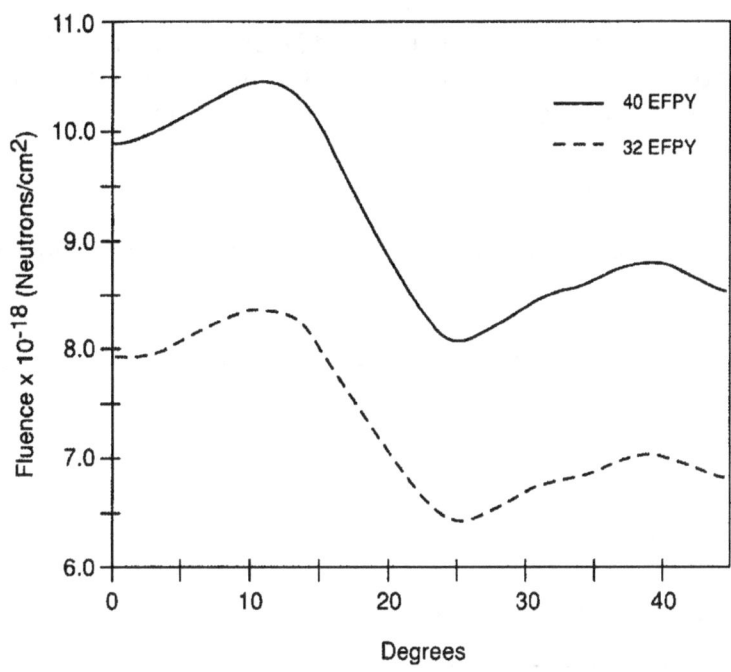

Figure 4.31 Azimuthal variation of fluence on the inner diameter of the Oconee 1 vessel at the axial location of the peak fluence

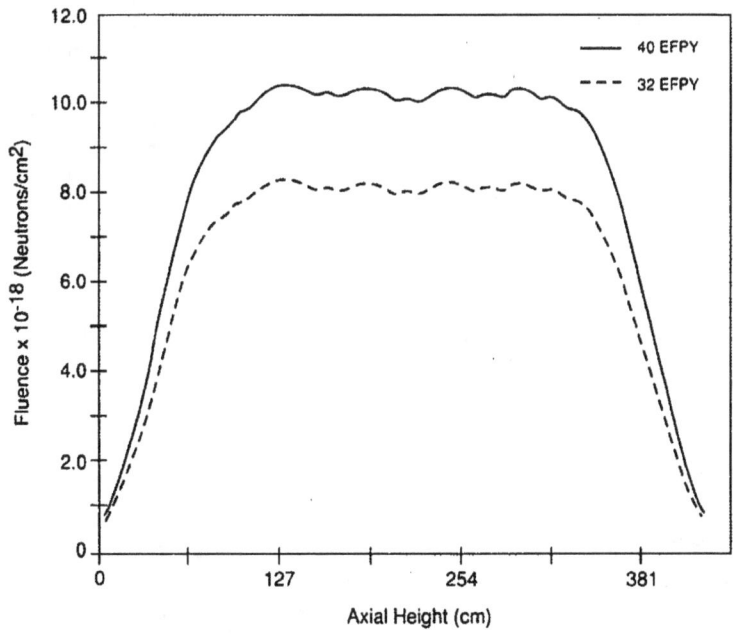

Figure 4.32 Axial variation of fluence on the inner diameter of the Oconee 1 vessel at the azimuthal location of peak fluence

These uncertainties manifest themselves at two different levels. All of the uncertainties listed can result in differences between the actual inside diameter fluence and the RG 1.190 estimate that are roughly equal over the entire vessel. However, some uncertainty sources (such as peripheral neutron source) can vary from location to location over the vessel wall. To model these effects, the FAVOR code includes simulation of fluence uncertainty in two places:

(1)   At the beginning of each vessel simulation, a difference value is sampled from a standard normal distribution having a standard deviation of 11.8 (called sampled value $\Delta\phi t_1$). The RG 1.190 fluence estimates (called ϕt) are then adjusted as follows:

$$\phi t_{new}\,(r,\theta,z) = \phi t_{1.190}(r,\theta,z)\cdot(1+\Delta\phi t_1/100)$$

These adjusted values are used for the remainder of this vessel simulation.

(2)   When each flaw is simulated, a difference value is sampled from a standard normal distribution having a standard deviation of 5.6 (called sampled value $\Delta\phi t_2$). The ϕt_{new} (r,θ,z) estimates are then adjusted as follows:

Eq. 4-8 $\phi t_{new}\,(r,\theta,z) = \phi t_{new}\,(r,\theta,z)\cdot(1+\Delta\phi t_2/100)$

These adjusted values are used for the remainder of this flaw simulation.

Table 4-4 Partitioning of Fluence Uncertainty (Values Based on Expert Opinion) (Carew 01)

Uncertainty Source	Origin of Uncertainty		Uncertainty Magnitude	
	Global (Vessel)	Local (Flaw)	Global (Vessel)	Local (Flaw)
Vessel Diameter	Mostly	Little	4.4%	0.0%
Peripheral Neutron Source	Yes	Yes	5.0%	5.0%
Core Inlet Temperature	Yes	No	1.8%	0.0%
Nuclear Cross Sections	Yes	No	7.0%	0.0%
Methods	Yes	No	6.0%	0.0%
Other	???	???	2.5%	2.5%
RMS Uncertainty			11.8%	5.6%

4.2.3.1.3   Attenuation Model and Uncertainty

Current information on how fluence attenuates through the wall of a thick reactor vessel is extremely limited. Regulatory Guide 1.99, "Radiation Embrittlement of Reactor Vessel Materials," Revision 2, adopts the following attenuation function:

Eq. 4-9 $\phi t(z) = \phi t \exp(-0.24z)$

where z is the distance from the inner diameter of the RPV. Eq. 4-9 assumes that fluence attenuates like displacements per atom, a conservative assumption (i.e., Eq. 4-9 assumes that fluence attenuates more slowly than it actually does). A recent review performed by the Electric Power Research Institute (EPRI) (see English 02) concluded that, while conservative, no better alternative estimates of attenuation exist than Eq. 4-9. For these reasons Eq. 4-9 is incorporated into FAVOR, and the effects of uncertainty on the relationship is not modeled.

4.2.3.2   Physics of Irradiation Damage of Ferritic RPV Steels

Neutron irradiation of U.S. RPV steels causes embrittlement effects marked by an increase in yield strength as a result of the fine scale microstructures produced by irradiation. These microstructures obstruct dislocation motion, thereby increasing the stress required to move dislocations past these obstacles. The following three mechanisms produce these obstacles:

(1) matrix hardening resulting from irradiation-produced point defect clusters and dislocation loops

(2) age hardening caused by irradiation-enhanced formation of copper-rich precipitates

(3) grain boundary segregation of embrittling elements such as phosphorus[‡‡]

Details of the two hardening forms of embrittlement are as follows:

(1) <u>Matrix hardening</u>. Matrix damage develops continuously during irradiation, producing hardening that has a square root dependence on fluence. Matrix damage can be divided into two components, unstable matrix defects (UMD) and stable matrix defects (SMD). UMDs are formed at relatively low fluence, and consist of small vacancy or interstitial clusters, complexed with solutes such as phosphorus and produced in displacement cascades. Increasing flux causes increasing hardening resulting from these defects, but they occur relatively independent of alloy composition. In low-copper alloys at low fluence and high flux UMD is the dominant source of hardening. However, in high-copper steels, these defects delay the copper-rich precipitate contribution to hardening by reducing the efficiency of radiation-enhanced diffusion. Stable matrix features form at high fluence and include nanovoids and more highly complexed clusters. These defects cause hardening that increases with the square root of exposure and is especially important at high fluences.

(2) <u>Age hardening</u>. Radiation accelerates the precipitation of copper held in solid solution, forming copper-rich precipitates that inhibit dislocation motion and, thereby, harden the material. This hardening rises to a peak value and is then unaffected by subsequent irradiation because no copper remains in solid solution to precipitate out and cause damage. The magnitude of this peak depends on the amount of copper initially in solution, and thereby available for subsequent precipitation. Postweld heat treatment (PWHT) performed before the RPV is placed into service can also precipitate copper, removing its ability to cause further damage during irradiation. Thus, different materials are expected to have different peak hardening values because of differing preservice thermal treatments. Additionally, the presence of nickel in the alloy further enhances its age-hardening capacity. Nickel precipitates together with copper, forming larger second-phase particles that present greater impediments to dislocation motion and thereby produce a greater hardening effect.

[‡‡] Irradiation can produce grain boundary segregation of tramp elements such as phosphorus. This leads to a nonhardening form of embrittlement (i.e., one that elevates the toughness transition temperature without increasing the yield strength). A broad technical consensus supports the notion that the steels in U.S. nuclear RPVs have sufficiently low impurity levels that nonhardening embrittlement is not expected. As the focus of this analysis is U.S. RPV steels, the physical understanding described in this section is appropriate only when hardening forms of embrittlement dominate. Application of these models to steels having higher impurity contents (e.g., VVER steels) is not appropriate.

The microstructures produced by both matrix and age hardening provide only long-range barriers to dislocation motion in BCC metals. The spacing of these irradiation-produced barriers occurs on a size scale many times larger than the lattice spacing of the atoms. Because the lattice spacing controls the temperature dependence of both the flow and toughness properties, the following effects of irradiation damage can be expected on a physical basis (see Natishan 01):

- Irradiation will increase the yield strength at all temperatures; it will not change the temperature dependency of the yield strength.

- Irradiation will shift the cleavage fracture toughness transition curve along the temperature axis; it will not change the temperature dependency of the cleavage fracture toughness.

Both of these physical expectations are supported by ample empirical evidence. Figure 4-9 already presented data showing that the physically anticipated effects of irradiation on cleavage fracture toughness manifest in reality (see Natishan 01). Similarly, data showing the physically anticipated effects of irradiation on yield strength are also available (see Figure 4-33 and Kirk 01a).

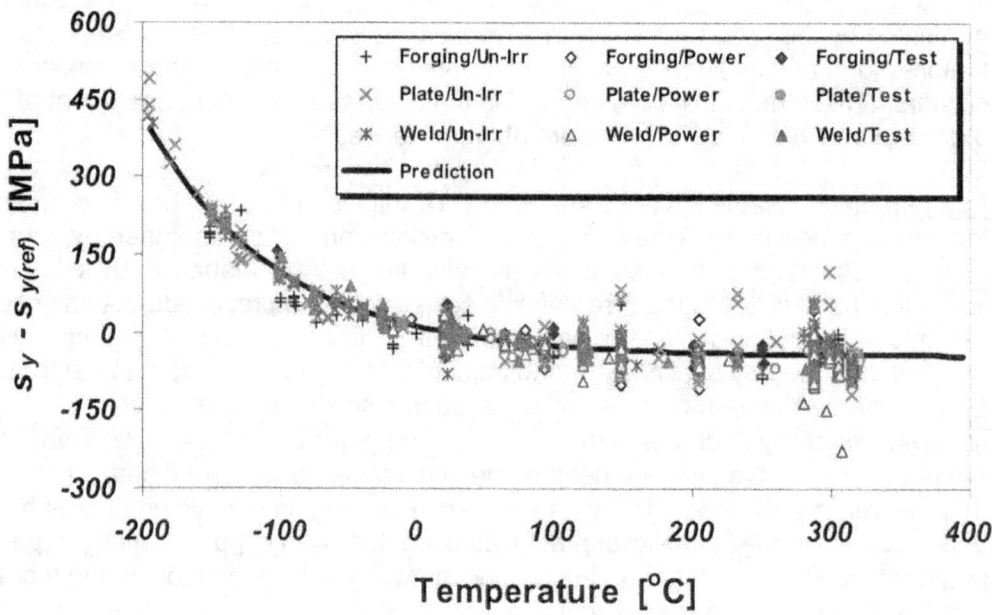

Figure 4-33 Comparison of 0.2 percent offset yield strength (s_y) for nuclear RPV steels to the Zerilli/Armstrong constitutive relation (line labeled "prediction") (Kirk 01a). $s_{y(ref)}$ is the ambient-temperature yield strength.

4.2.3.3   Charpy Irradiation Shift Model

4.2.3.3.1 Model Selected

The most comprehensive models available today concerning the effects of neutron irradiation on the mechanical properties of ferritic RPV steels relate basic compositional and irradiation variables to the shift in CVN energy transition temperature (ΔT_{30}), rather than to shifts in the toughness transition temperature. This focus on ΔT_{30} results from the historical practice of measuring Charpy shift as part of RPV surveillance programs, making the great preponderance of the data available for calibrating irradiation-effects model Charpy data, rather than fracture toughness data. Thus, while Charpy specimens do not measure fracture toughness, and while Charpy data exhibit trends that are not physically justifiable (i.e., irradiation changes the temperature dependency of Charpy energy, and Charpy shifts exhibit a dependence on product form), matters of expedience dictate the use of Charpy-based irradiation-shift models (with a subsequent conversion of Charpy shift to toughness shift). This section discusses the Charpy-shift model adopted for FAVOR, reserving discussion of the conversion of Charpy shift to toughness shift for the following section.

The Charpy embrittlement model adopted for FAVOR follows that proposed by Eason (see Eason 03). Figure 4-34 illustrates the algebraic form of this model, along with the calibration data set (all available data from U.S. power reactor surveillance programs docketed with the NRC by 1999), and Table 4-5 provides the units for the independent variables. While the numerical coefficients in this model were determined by nonlinear least squares fitting to these data, the model is physically motivated in the sense that the algebraic forms selected for fitting derive, in many cases, from the following physical understanding of the physics of irradiation damage (see Section 4.2.3.1.3):

- different (additive) terms to reflect the different nature of the physical contributions of matrix hardening ("A" term) and age hardening ("B" term) described in Section 4.2.3.1.3

- in the matrix hardening (A) term:

    — a (nearly) square root dependency on fluence
    — a dependency on phosphorus and an independence from other embrittling elements

- in the age hardening (B) term:

    — a threshold copper level below which no age hardening occurs, leaving the matrix hardening term to completely dominate the irradiation response for low-copper alloys
    — saturation in age hardening at high copper levels that, through the use of flux type as an indicator variable, corresponds to differences in PWHT practice
    — a synergistic effect between copper and nickel that leads to greater hardening

Eason's equation also includes a number of features that rely more heavily on either an empirical understanding of irradiation effects, or on a recently emerging physical understanding, including the following items:

- a synergistic effect of flux and time
- a product form dependency (including an effect of vessel manufacturer)
- a purely time-dependent effect

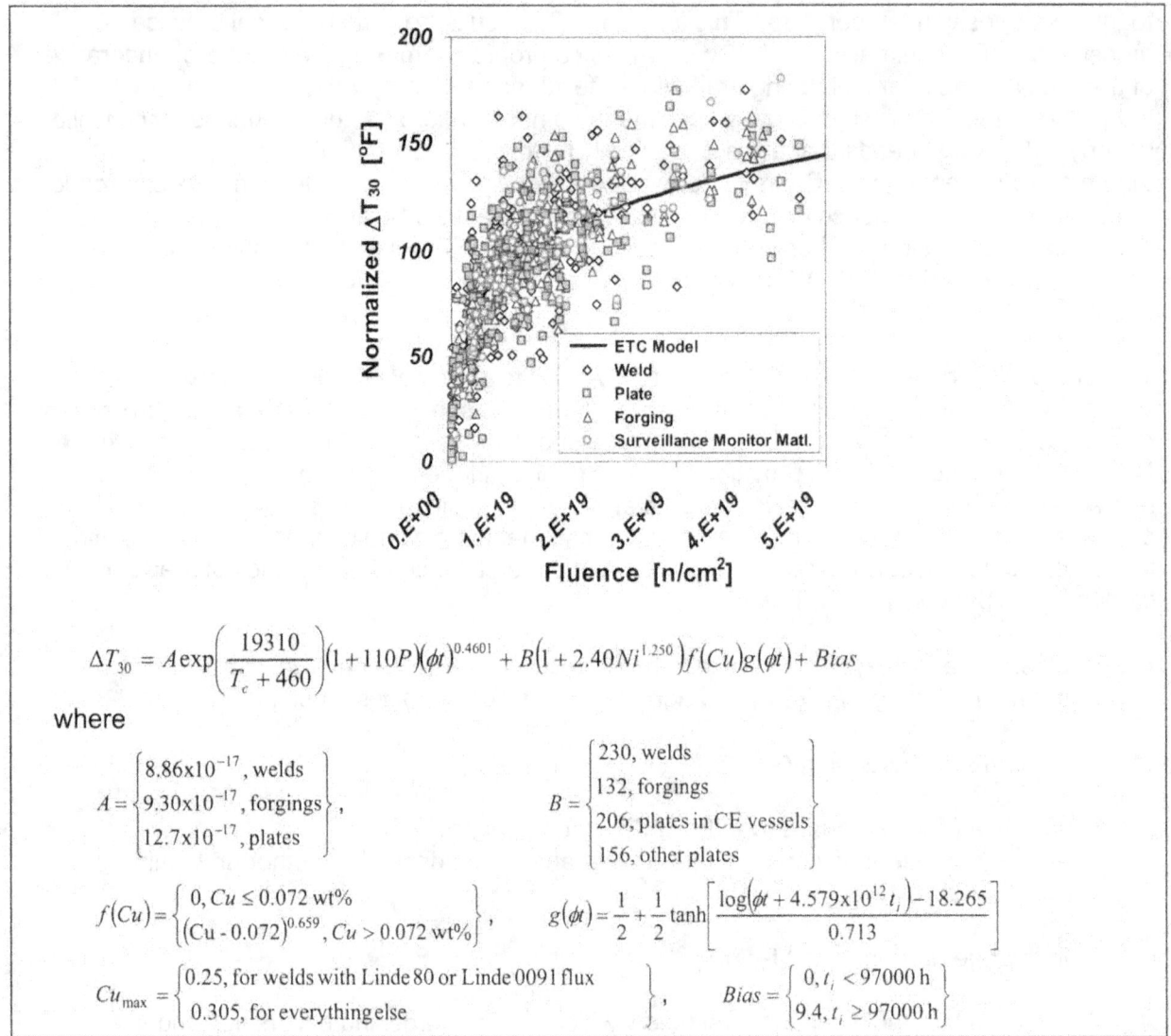

$$\Delta T_{30} = A\exp\left(\frac{19310}{T_c + 460}\right)\left(1 + 110P\right)\left(\phi t\right)^{0.4601} + B\left(1 + 2.40Ni^{1.250}\right)f\left(Cu\right)g\left(\phi t\right) + Bias$$

where

$$A = \begin{cases} 8.86\text{x}10^{-17}, \text{welds} \\ 9.30\text{x}10^{-17}, \text{forgings} \\ 12.7\text{x}10^{-17}, \text{plates} \end{cases}, \qquad B = \begin{cases} 230, \text{welds} \\ 132, \text{forgings} \\ 206, \text{plates in CE vessels} \\ 156, \text{other plates} \end{cases}$$

$$f\left(Cu\right) = \begin{cases} 0, Cu \le 0.072 \text{ wt}\% \\ \left(Cu - 0.072\right)^{0.659}, Cu > 0.072 \text{ wt}\% \end{cases}, \qquad g\left(\phi t\right) = \frac{1}{2} + \frac{1}{2}\tanh\left[\frac{\log\left(\phi t + 4.579\text{x}10^{12}t_i\right) - 18.265}{0.713}\right]$$

$$Cu_{max} = \begin{cases} 0.25, \text{for welds with Linde 80 or Linde 0091 flux} \\ 0.305, \text{for everything else} \end{cases}, \qquad Bias = \begin{cases} 0, t_i < 97000 \text{ h} \\ 9.4, t_i \ge 97000 \text{ h} \end{cases}$$

Figure 4-34 Eason embrittlement trend curve model

Table 4-5 Independent Variables in Figure 4-34

Variable	Description	Range (Calibration)	Median	Units
Cu	Copper content	0.01–0.42	0.133	wt%
Ni	Nickel content	0.044–1.26	0.6	wt%
P	Phosphorus content	0.003–0.031	0.011	wt%
ϕt	Neutron fluence	$9.26\text{x}10^{15}$–$1.07\text{x}10^{20}$	$8.66\text{x}10^{18}$	n/cm^2, E>1MeV
t_i	Exposure time	5556–158,840	38,025	Hours
T_c	Coolant temperature	522– 570	545	°F

The ASTM recently adopted a new Charpy irradiation-shift model as part of ASTM Standard Guide E900-02. The ASTM model omits the last three features incorporated by the Eason model. The effect of these model differences on the value of through-wall cracking frequency (TWCF) estimated by FAVOR is addressed in a separate report on sensitivity studies (see EricksonKirk 04c).

4.2.3.3.2 Uncertainty Treatment

The comparison of "normalized" ΔT_{30} values[§§] to the prediction of Eason's model shown in Figure 4-34 demonstrates that there is considerable scatter about the mean predicted value, even for the data used to develop the model. The following two sources of uncertainty contribute to the observed scatter:

(1)   how well the mathematical form of the Charpy embrittlement model represents the physical processes of irradiation damage

(2)   how accurately and consistently the data set used to calibrate the embrittlement trend curve represent the data (i.e., ΔT_{30} data, chemical composition data, and fluence data)

Both of these uncertainties relate to a lack of knowledge, and so both are (primarily) epistemic in nature. FAVOR does not model the first uncertainty because no other available model (ASTM E900-02 or otherwise) exhibits appreciably less uncertainty. Thus, there is no independent metric of truth relative to which this uncertainty could be quantified (and thereby modeled).

The second source of uncertainty (uncertainty in the measured values and distributions of the input parameters to the Charpy embrittlement relationship) represents the major source of uncertainty that can be quantified and therefore modeled in FAVOR. As reflected in Figure 4-30, the NRC uncertainty model involves using the copper, nickel, and phosphorous values taken from RVID (see Appendix C) to center generic copper, nickel, and phosphorous distributions that are developed in Appendix D, as well as a Monte Carlo process to draw individual samples from these distributions. These individual samples of copper, nickel, phosphorous, and fluence (see Section 4.2.3.1 for a discussion of the treatment of fluence uncertainties) are then propagated through FAVOR in a particular simulation run.

It should be noted that the FAVOR procedure does not also simulate the uncertainty in the Charpy embrittlement model shown on Figure 4-34. This approach is appropriate because the uncertainty in the embrittlement shift model arises from uncertainties in the input variables to the embrittlement shift model (i.e., copper content, nickel content, phosphorus content, and fluence) which are sampled in FAVOR. This is demonstrated by the results in Figure 4-35, which were generated as follows:

(1)   Median values were assigned to all of the input variables to the Eason embrittlement shift equation (except for fluence).

[§§] "Normalized" ΔT_{30} values are the measured ΔT_{30} values normalized to median values of the independent variables (see Table 4-5) to simulate the appearance of data as if all ΔT_{30} values were determined from tests conducted under the same conditions.

(2)   The FAVOR uncertainty distributions on copper, nickel, phosphorus, and fluence were sampled about these medians for fluence medians ranging from 0.25×10^{19} to 5×10^{19} n/cm^2.

(3)   At each different fluence value, 1000 sets of copper, nickel, phosphorous, and fluence data were simulated. Each set was used to estimate a value of embrittlement shift using the Eason embrittlement model. The standard deviation of these 1000 embrittlement shift estimates was calculated and plotted in Figure 4-35.

The uncertainties simulated by FAVOR agree well with the uncertainties in the embrittlement shift data used by Eason to develop the model. The lower uncertainties associated with lower fluence values results from FAVOR setting to zero simulations of embrittlement shift that are negative, which is physically unrealistic.

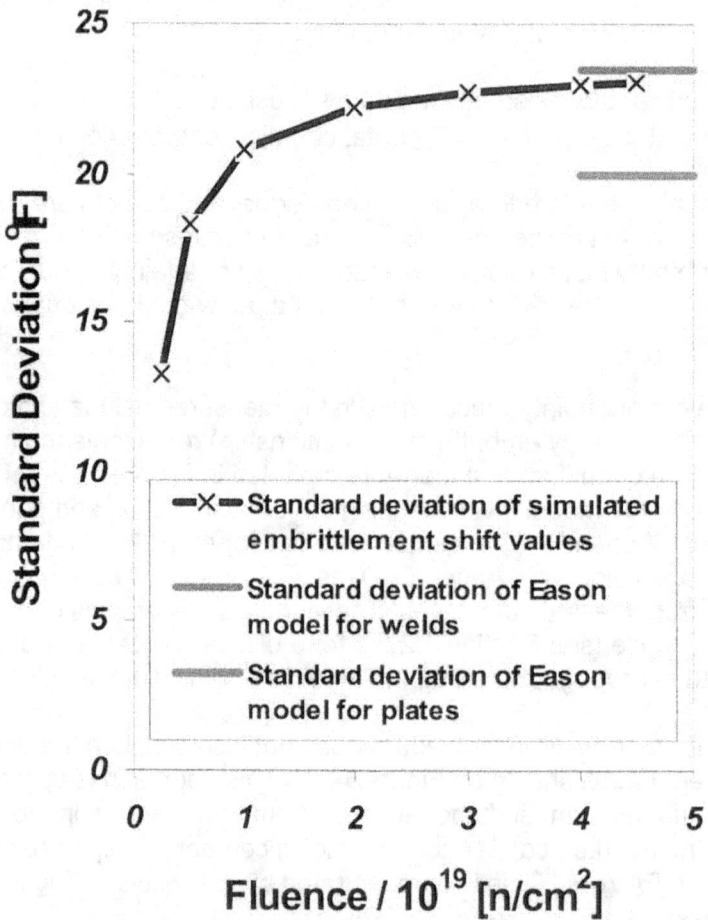

Figure 4-35 Comparison of embrittlement shift uncertainties simulated by FAVOR (blue line with X symbols) with the uncertainties in the experimental embrittlement shift database used by Eason to construct the model

This information confirms the appropriateness of this approach to uncertainty simulation for the model. Simulation of both the embrittlement shift model uncertainties and the uncertainties in the input variables would produce a model that simulated a greater magnitude of uncertainty in embrittlement shift than is observed in test data.

4.2.3.4 Conversion of Charpy Shift to Toughness Shift

4.2.3.4.1 Model Selected

The understanding of the mechanisms of irradiation damage summarized in Section 4.2.3.1.3 suggests that the increase in the room temperature yield strength (Δs_{ys}) produced by irradiation provides a physically motivated quantification of the degree of irradiation damage imparted to a ferritic steel. Measured values of Δs_{ys} therefore include the complex irradiation mechanics that the empirically derived embrittlement trend curves attempt to capture. Figure 4-36 and Figure 4-37 therefore examine the relationships of both ΔT_o and ΔT_{41J} to Δs_{ys} using available empirical data. The linear trend clearly evident in both of these figures is consistent with the idea introduced in Section 4.2.3.1.3 that the contributions of both matrix and age-hardening embrittlement can be captured as an addition to the athermal part of the flow stress. Further examination of these data reveals that a product-form dependence not evident in the transition fracture toughness data (i.e., in ΔT_o) appears when transition is described using Charpy data (i.e., ΔT_{41J}). This dependency, which also appears in the ΔT_{41J} Charpy embrittlement model (see Figure 4-34), is not expected on physical grounds. Differences in thermomechanical processing related to the product form that clearly influence absolute transition temperatures play no role in transition temperature shift values because the physical basis for product-form effects influences unirradiated and irradiated transition temperature values equally, resulting in their cancellation when transition temperature shifts are calculated. Thus, the trends exhibited by the ΔT_o data in Figure 4-36 are anticipated physically, while the trends exhibited by the ΔT_{41J} in Figure 4-37 are not.

Some rationalization of product form-dependent ΔT_{41J} values can be obtained from a more detailed examination of CVN data. While the shape of the fracture toughness transition curve (i.e., the Master Curve) is invariant with irradiation and product form (see Wallin and Natishan), the shape of the CVN transition curve is not. As illustrated in Figure 4-38, this results in a 41J transition temperature being defined at different locations on the CVN transition curve for different product forms (specifically, after irradiation 41J is considerably closer to the upper-shelf energy (USE) for welds than it is for either plates or forgings). As the USE approaches the 41J level, the CVN transition curve lays over, which results in a progressively greater degree of CVN transition temperature shift for the same irradiation and chemistry conditions. While simplified, this explanation rationalizes the existence of product form dependencies in ΔT_{41J}, but not in ΔT_o. Additionally, the explanation successfully ranks welds, plates, and forgings in terms of the amount of CVN transition temperature shift that can be expected for a fixed severity of irradiation damage (fixed Δs_{ys}).

When the ΔT_o data from Figure 4-36 are plotted against the corresponding ΔT_{41J} values from Figure 4-37, a product from dependent correlation results as a direct consequence of the product-form dependencies in ΔT_{41J} just discussed (see Figure 4-39). These data show that at a fixed value of ΔT_{41J}, progressively greater ΔT_o shifts are expected from (in order) welds, then plates, and then forgings, specifically as follows:

Eq. 4-10 $\quad \Delta T_o = \alpha \cdot \Delta T_{30}$

where α is 0.99 for welds, 1.10 for plates, and 1.50 for forgings (the forging α value is based on exceedingly limited data and should be treated with caution[***]). However, the α-values in Eq. 4-10, which rank product forms (in terms of increasing irradiation shift severity) as weld-plate-forging serve to counteract the B-coefficients in the Charpy embrittlement model (Figure 4-34), which rank product forms (in terms of increasing irradiation shift severity) as forging-weld-plate. Consequently, this method of estimating the irradiation-induced shift in ΔT_o from ΔT_{41J} or ΔT_{30} merely restores (at least approximately) the physically expected product form insensitivity of toughness shift values.

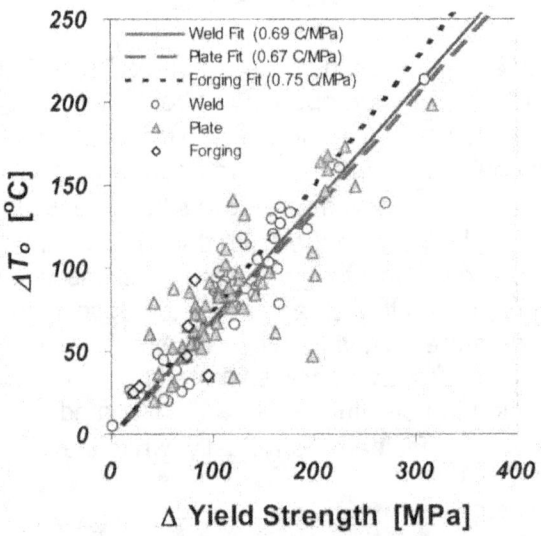

Figure 4-36 Relationship between the change in the fracture toughness index temperature (ΔT_o) and the increase in the room temperature yield strength produced by irradiation. The differences in the fit slopes are not statistically significant.

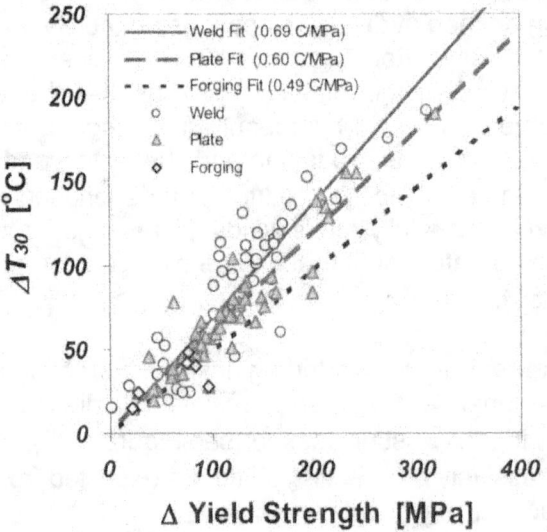

Figure 4-37 Relationship between the change in the 41J ft-lb CVN transition temperature (ΔT_{41J}) and the increase in the room temperature yield strength produced by irradiation. The differences in the fit slopes are statistically significant.

[***] In FAVOR the α-coefficient for forgings is set to the plate value of 1.1 because of the limited amount of data available for forgings.

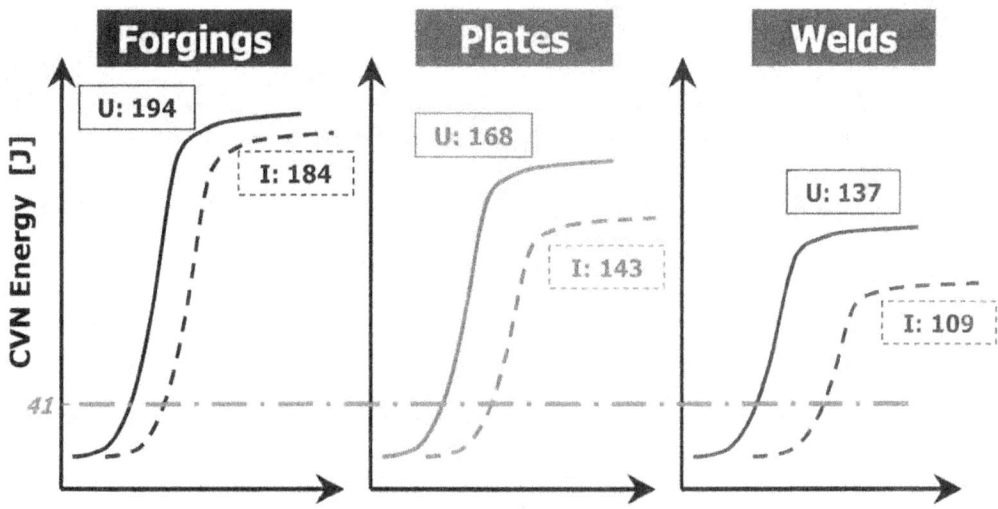

Figure 4-38 Illustration of how differences in USE can influence the 41J transition temperature. The USE given in the boxes for (U) unirradiated and (I) irradiated conditions are the averages determined from the data used to calibrate the Charpy embrittlement model shown in Figure 4-34.

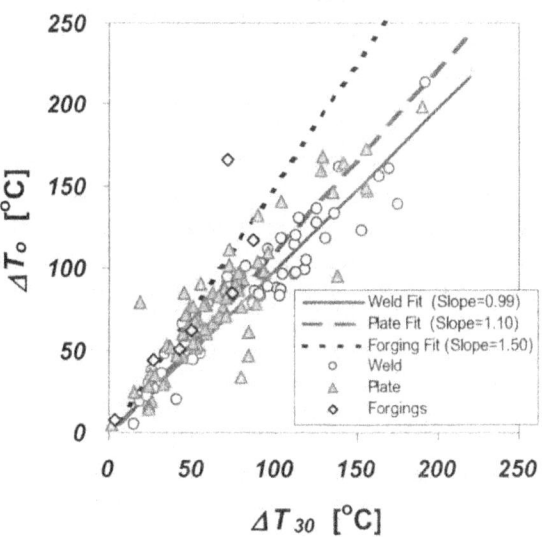

Figure 4-39 Relationship between the change in the fracture toughness index temperature (ΔT_o) and the change in the 30 ft-lb CVN transition temperature (ΔT_{41J}) produced by irradiation. The differences in the fit slopes are statistically significant.

4.2.3.4.2   Uncertainty Treatment

FAVOR represents the model for conversion to toughness shift (Eq. 4-10 and Figure 4-39) as being without uncertainty for the following reasons:

(1)   When attention is restricted to the better-defined values of ΔT_o and ΔT_{30}, considerably less scatter is seen in the relationship (see Figure 4-40), suggesting that the uncertainty shown in Figure 4-39 depends much more on measurement uncertainty than uncertainty in the physical processes underlying the relationship.

(2)   It is well established from experimental evidence that the variability in ΔT_o and ΔT_{30} are of approximately equal magnitude. However, as illustrated in Figure 4-41, incorporating the variability in the ΔT_o vs. ΔT_{30} experimental relationship produces a considerable (and unrealistic) broadening in the range of predicted ΔT_o values.

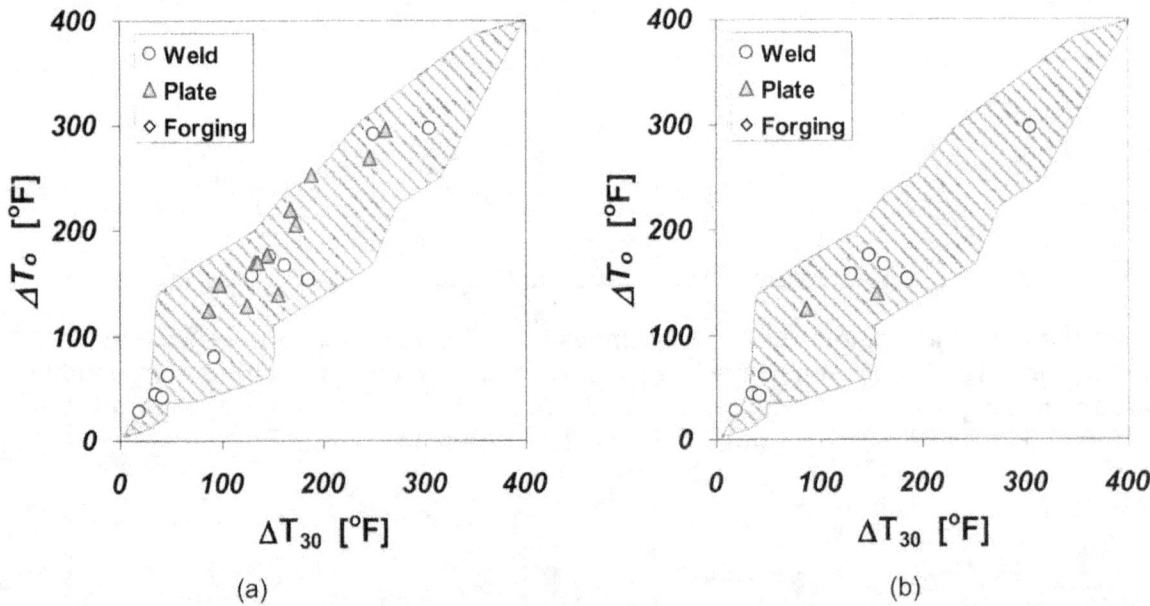

Figure 4-40 Relationship between CVN shift caused by irradiation (ΔT_{30}) and the shift in the fracture toughness transition temperature caused by irradiation T_o (ΔT_o) for data sets limited to (a) 10 K_{Jc} values or more, and (b) 15 K_{Jc} values or more. Shaded region shows the range of all the data from Figure 4-39 (all data determined with 6 K_{Jc} values or more).

4.2.3.5   Summary of Model and Uncertainty Treatment

Figure 4-33 combines the diagrammatic representation of the index temperature shift model taken from Figure 4-30 with the specific models and input values the preceding sections recommend for use. Uncertainties are accounted for in the index temperature shift model in the following two places:

- Fluence uncertainty is modeled (see Section 4.2.3.1.2).

- Uncertainty in chemical composition is modeled (see Appendix D). Also, it is important to point out that narrower uncertainty distributions are sampled the second (and third, and fourth, etc.) time a flaw is simulated to reside in a particular subregion in a particular vessel than are sampled the first time a flaw is simulated to reside in a particular subregion. This approach reflects the idea that composition does not vary as much over small volumes of material as it does over large volumes of material, an idea supported by experimental evidence.

Of equal importance, Figure 4-33 reflects the idea that in three places the index temperature shift model should not include the following uncertainties:

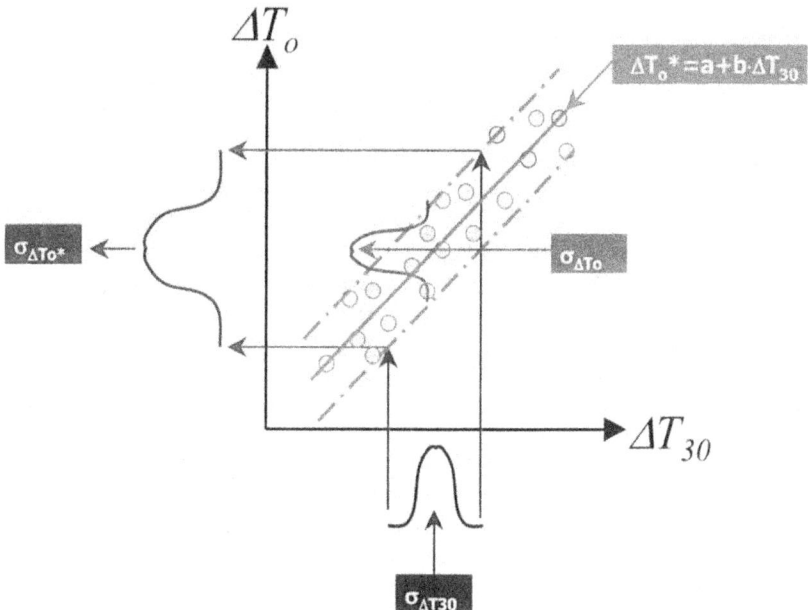

Figure 4-41 Illustration of how the uncertainty in predicted values of ΔT_o is broadened when uncertainty in the empirical ΔT_o vs. ΔT_{30} relationship is included in the calculation

FAVOR represents all of the above uncertainties as being epistemic. Thus, individual values of the uncertain variables are used throughout a simulation run, after which they are resampled.

- Uncertainties are not included in the fluence attenuation model because the model is conservative, and no alternative model exists that could be used either (1) in its place, or (2) to assess its conservatism.

- Uncertainties are not included in the Charpy irradiation shift model because the apparent uncertainties in the model have already been accounted for in the process of simulating the uncertainties in the compositional variables and in fluence. Including uncertainties in the Charpy irradiation shift model would therefore result in an inappropriate double counting of uncertainties, and consequent simulation of Charpy shift values that are more uncertain than reflected by available empirical evidence.

- Uncertainties are not included in the model that converts Charpy shift to toughness shift because doing so would result in a simulation of toughness shift values that are more uncertain than those reflected by available empirical evidence.

4.2.4   Interface Model

In a destructive examination of flaws in RPV welds, Simonen reported that virtually all flaws in nuclear-grade welds (95 percent or greater) are found on the fusion line between the weld metal and the adjacent base metal (plate or forging). The FAVOR crack initiation model assumes that the fracture toughness transition reference temperature that characterizes the material at the crack tip is the maximum of fracture toughness transition reference temperature of the plate material that exists on one side of the flaw and the weld material that exists on the other. This model assumes that if a crack propagates it will do so preferentially through the most brittle material available. This model is implemented without uncertainty.

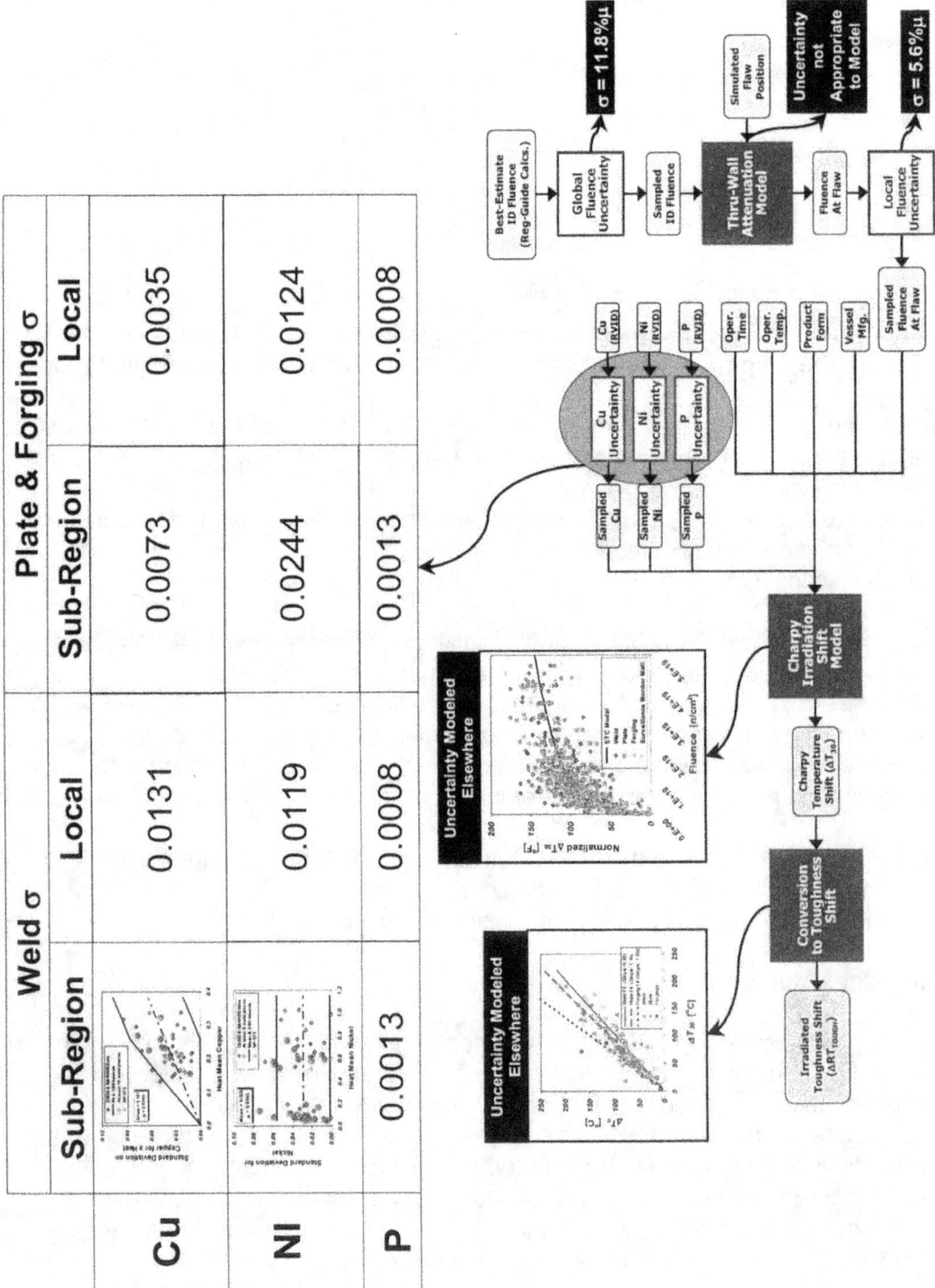

Figure 4-42 Diagrammatic summary of the FAVOR embrittlement model showing the recommended models, input values, and uncertainty treatment

4-52

5   THROUGH-WALL CRACKING MODEL

The model of through-wall cracking adopted for FAVOR compares the applied driving force to fracture and the material's resistance to further cracking to estimate the conditional probability of through-wall cracking. The model of the applied driving force to fracture is the same as the linear elastic fracture mechanics (LEFM) model used for the crack initiation calculation (see Section 4.1.1). In estimating the material's resistance to further cracking, the U.S. Nuclear Regulatory Commission (NRC) admits the possibility, after crack arrest, of reinitiation in either a brittle mode, which is controlled by the K_{Ic}/RT_{NDT} model discussed in Chapter 4, or in a ductile mode, which is controlled by the material's J_{Ic} and J-R curve properties[†††].

As illustrated in Figure 5-1, the through-wall cracking model is composed of the following parts:

- a fracture driving force model
- a fracture resistance model, which includes:

  — the assumption that upon crack initiation the crack immediately grows along the inner diameter surface of the vessel to a length that greatly exceeds its through-wall depth

  — a model that describes the gradient of material properties through the vessel wall thickness

  — a crack arrest model

  — a ductile tearing model

Several aspects of these models have been discussed previously:

- The fracture driving force model used in this analysis is identical to the LEFM model, as was used for the crack initiation calculation (see Section 4.1.1).

- The index temperature shift model discussed in Section 4.2.3 forms part of the crack arrest model.

- Appendix A summarizes the experimental observations justifying the assumption that upon crack initiation the crack immediately grows along the inner diameter surface of the vessel to a length that greatly exceeds its through-wall depth.

For a discussion of these aspects of the through-wall cracking model the reader should refer to the relevant sections of this report. The remainder of this chapter concerns the following topics:

- Section 5.1 describes the crack arrest toughness model.

[†††] Inclusion of ductile fracture properties in a probabilistic pressurized thermal shock (PTS) model is a departure from previous calculations, which considered only brittle fracture properties [SECY-82-465, ORNL 85a, ORNL 85b, ORNL 86]. Section 5.2 discusses the need to consider tearing on the upper shelf as a potential failure mode in PTS calculations.

- Section 5.2 describes the upper-shelf ductile tearing model.
- Section 5.3 describes the material property gradient model.

5.1 Crack Arrest Toughness Model

5.1.1 The Physics of Crack Arrest in Ferritic Steels

The information presented in Section 4.2.1 included a description of the physics of cleavage fracture in ferritic steels, and established the physical basis for both a temperature dependency and a scatter in crack arrest toughness that is universal to all ferritic steels. These physical expectations will factor into the proposed model for FAVOR, which Section 5.1.2 describes. In addition to temperature dependency and scatter, the crack arrest model should also reflect the relationship between crack initiation and crack arrest index temperatures. The remainder of this section describes the physical basis for such a relationship.

Wagenhofer and Natishan (see Wagenhofer 01) contend that a universal hardening curve exists for all ferritic steels. This section summarizes this idea, and then develops a physical model for the relationship between the crack arrest and crack initiation transition index temperatures using the universal hardening curve proposed by Wagenhofer and Natishan as a basis.

The following Zerilli-Armstrong description of the flow curve for BCC materials explains the relationships between the various parameters involved (see Zerilli 87):

Eq. 5-1
$$\sigma_{ZA} = \sigma_0 + C\varepsilon^n \qquad \text{(a)}$$
$$\sigma_0 = c_0 + Be^{-\beta T} \qquad \text{(b)}$$
$$c_0 = \sigma_G + kd^{-1/2} \qquad \text{(c)}$$
$$\beta = \beta_0 - \beta_1 \ln\dot{\varepsilon} \qquad \text{(d)}$$

Here, K, n, c_0, B, k, β_0 and β_1 are material constants, σ_G is the increment of true stress resulting from coherent and semicoherent obstacles to dislocation motion (vacancies, interstitials, and small precipitates), ε is the true strain, $\dot{\varepsilon}$ is the true strain rate, and d is the grain size. Focusing attention on the work-hardening term (2nd term on right side of Eq. 5-1(a)), the notion that all ferritic steels follow a universal hardening curve (i.e., the same strain-hardening rate, n, value for all ferritic steels) follows directly from the well-established experimental observation that as a metal is hardened, subsequent tensile tests will reveal greater yield strengths, but the overall true stress-strain curve will always overlay the unhardened curve. This behavior cannot occur unless the hardened specimen exhibits the same strain-hardening rate as the unhardened specimen does after an equivalent amount of tensile strain. This observation also leads to an invariance of the true stress at maximum load for most hardening mechanisms.

In Eq. 5-1, Zerilli and Armstrong adopt $n=0.5$, following Taylor's derivation of the stress necessary to keep a uniform distribution of edge dislocations in equilibrium (see Taylor 34). This square root relationship between stress and strain arises because the uniform dislocation distribution forces the distance between the dislocations to be equal to the inverse square root of the dislocation density. Despite criticisms that this theory is too simplistic, it does a good job of representing the large deformation behavior of a number of materials (see Meyers 99).

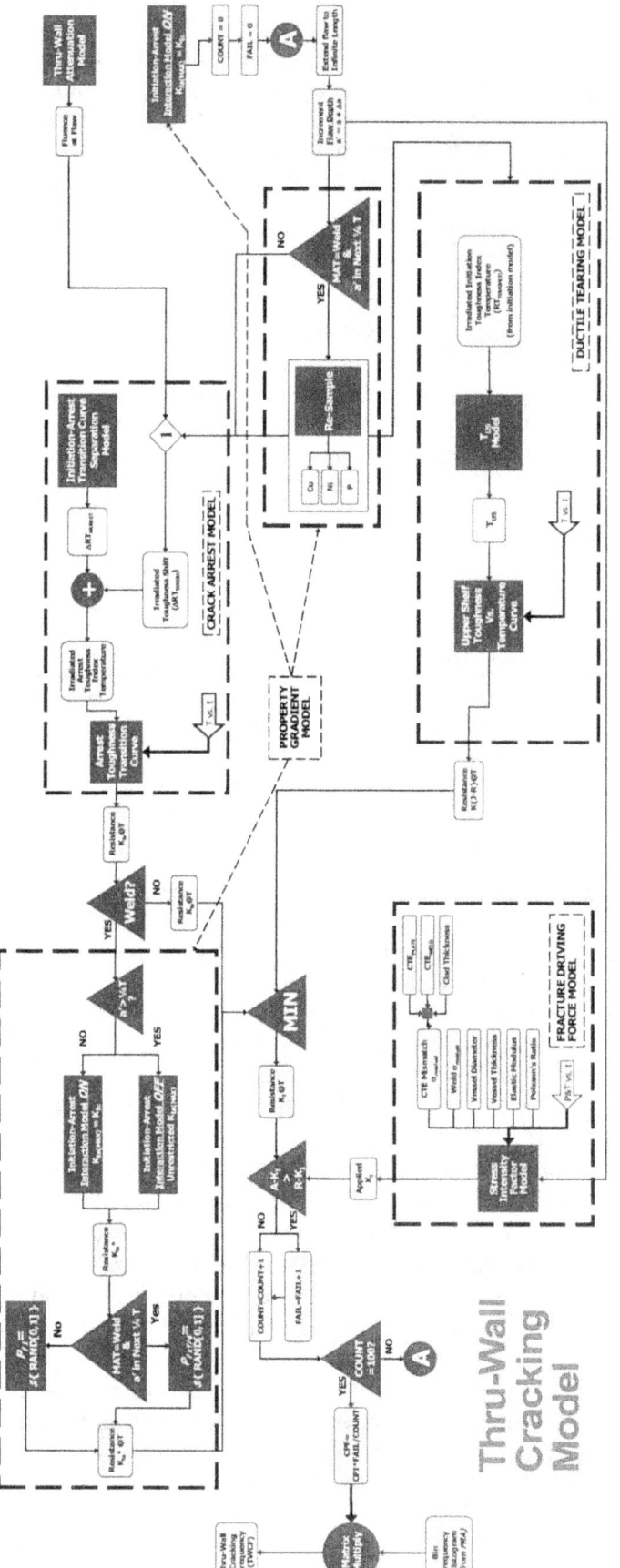

Figure 5-1 FAVOR through-wall cracking model

To summarize, for Eq. 5-1 to properly describe the physical behavior of metals that have been hardened to some degree, the strain hardening rate, n, must be constant, and a value of n equal to 0.5 is expected on theoretical grounds. The stress values for a particular steel having a particular degree of prior hardening can be determined by modifying Eq. 5-1(a) as follows (see Datsco 66):

Eq. 5-2 $\sigma_{ZA} = \sigma_0 + C\sqrt{(\varepsilon_0 + \varepsilon)}$

where ε_0 is a constant that quantifies the degree of prior hardening.

Figure 5-2 illustrates the effect of prior hardening on ARMCO Iron (see Zerilli 87) using Eq. 5-1. The thin solid lines are the engineering stress vs. true plastic strain curves for various amounts of prior hardening (various ε_0 in Eq. 5-2). These are calculated from Eq. 5-1 (thick solid line) using the familiar relationships:

Eq. 5-3 $\quad \sigma_{ZA} = S(1+e)$ $\qquad$ (a)

$\quad \varepsilon = \ln(1+e)$ $\qquad$ (b)

Here S and e are engineering stress and strain, respectively. Various amounts of prior hardening are represented as initial tensile strains (thin vertical lines on Figure 5-2). The maximum load condition is as follows:

Eq. 5-4 $\quad \dfrac{d\sigma_{ZA}}{d\varepsilon} = \sigma_{ZA}$,

This is represented on Figure 5-2 by the dotted line.

These ideas provide the basis for a physical model of the relationship between crack arrest and the crack initiation transition temperature reported by Wallin, as shown in Figure 5-4 (see Wallin 98a). At the time of crack arrest, the material experiences a high rate of loading. This loading-rate elevation above the quasi-static conditions associated with crack initiation causes an elevation in the activation energy required to move dislocations past trapping obstacles, and thus results in an increase in apparent yield stress of the material in a manner similar to the yield stress elevation produced by prior strain illustrated in Figure 5-2. Figure 5-3 uses the idea of a universal hardening curve for all ferritic steels to illustrate why the elevation in prior strain caused by the elevated loading rate associated with crack arrest (defined as $\Delta\varepsilon_0$) produces a progressively diminishing elevation in the yield strength as the degree of strain caused by prior hardening (ε_0) increases. Since increases in transition temperature scale with increases in yield strength (see Kirk 01b), this understanding suggests a physical basis for the empirical trend reported by Wallin of a progressively diminishing separation between the crack initiation and crack arrest transition curves for higher transition temperature steels. Moreover, the invariance of the true stress at maximum load that follows directly from the notion of a universal hardening curve suggests that, in the limit of very high strength ferritic materials, the crack initiation and crack arrest transition curves should approach each other (i.e., T_{KIa} is approximately equal to T_o), a trend also reflected by available empirical evidence (see Figure 5-4).

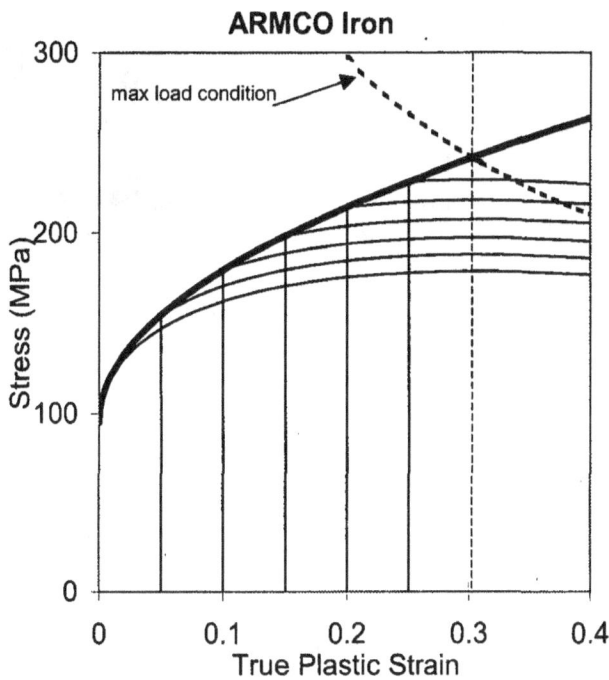

Figure 5-2 Engineering stress vs. true plastic strain curves for various degrees of prior hardening (thin vertical lines and curves) calculated from the Zerilli/Armstrong true stress vs. true plastic strain curve (thick curve) for ARMCO Iron

5.1.2 Proposed Model

5.1.2.1 K_{Ia} Index Temperature

As suggested by Wallin (and illustrated in Figure 4-10), the NRC adopted T_{KIa} as a consistent crack arrest index temperature. Because T_{KIa} is defined as the temperature at which the mean crack arrest toughness is equal to 100 MPa√m, it must always correspond to the location of the crack arrest toughness transition data.

5.1.2.2 Temperature Dependency of K_{Ia}

Section 4.2.1 argued, on physical grounds, that all ferritic steels should share a common temperature dependency of K_{Ia}. Wallin has argued empirically that K_{Ia} data follow the same trends as the crack initiation toughness data (i.e., they follow the Master Curve; see Wallin 97). Kirk and Natishan present a physical argument for a slight difference between the temperature dependency of crack arrest and crack initiation toughness as a consequence of the different loading rates associated with the two different events (initiation vs. arrest) (see Kirk 02).

For FAVOR, the NRC developed empirical fits to crack arrest data indexed to T_{KIa}. Two crack-arrest transition models were developed, one fit to only E1221 K_{Ia} data (see Figure 5-5) and one fit to both E1221 data, as well as to wide-plate data, thermal shock data, and PTS data (see Figure 5-6). Over the temperature ranges where the models overlap, the fits are similar. The second model (Figure 5-6) is used in calculations that support the PTS reevaluation effort. The high K_{Ia}/high arrest temperature data reflected in this model are necessary to properly characterize the behavior of cracks as they approach the outer diameter of the reactor pressure vessel (RPV).

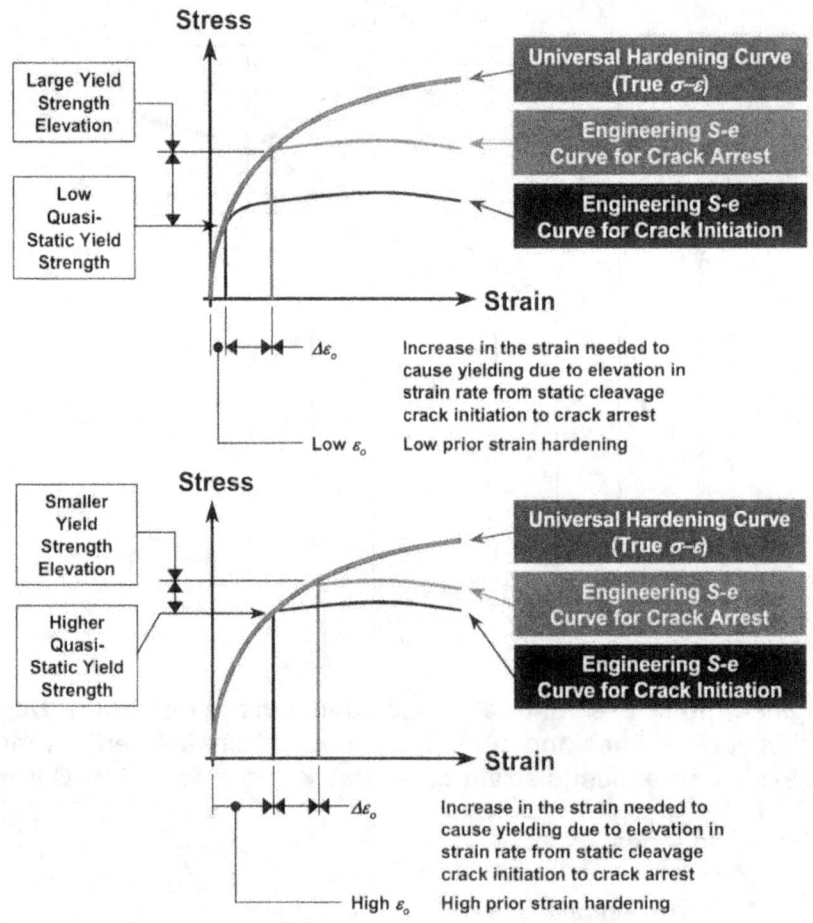

Figure 5-3 An illustration of the effect of strain rate increase on yield-strength elevation for materials having different degrees of prior strain hardening

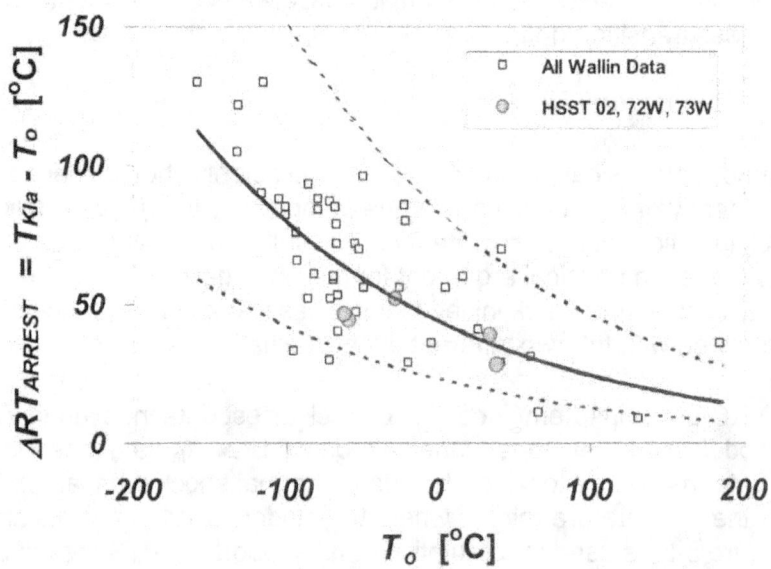

Figure 5-4 Data for RPV and other steels showing the separation between the crack arrest transition temperature ($\Delta RT_{ARREST} = 44.1 \cdot \exp[-0.006 \cdot T_o]$) (Wallin 98b)

Figure 5-7 compares the temperature dependencies of the FAVOR K_{Ic} model, the two FAVOR K_{Ia} models, and the Wallin Master Curve. The temperature dependency of all these models is quite similar, in agreement with expectations premised on physical considerations.

5.1.2.3   Relationship of K_{Ic} and K_{Ia} Transition Curves

The NRC adopted the relationship illustrated in Figure 5-4 to define the temperature separation between the crack initiation toughness and the crack arrest toughness transition curves. However, this relationship cannot be used directly because the index temperature adopted in the K_{Ic} model of Eq. 4-7 uses RT_{LB} as the index temperature, rather than T_o. The similarity of the T_o- and RT_{LB}-based cumulative distribution functions (CDFs) depicted in Figure 4-27 was therefore used to relate RT_{LB} to T_o. Figure 4-27(b) suggests that, on average,

Eq. 5-5 $T_o = RT_{LB} - 14.4$ (in °F)

Substituting Eq. 5-5 into the relationship shown on Figure 5-4 gives the following:

Eq. 5-6 $\Delta RT_{ARREST} \equiv 44.1 \cdot \exp\{-0.006(RT_{LB} - 14.4 - 32)/1.8\}$

Eq. 5-6 now describes the separation between a K_{Ic} curve (indexed to RT_{LB}) and a K_{Ia} curve (indexed to T_{KIa}). The uncertainty on this estimate is described by a log-normal distribution having an estimated standard deviation of 0.39.

5.1.2.4 Model Summary, Uncertainty Classification, and Treatment

Uncertainties occur in two places in the crack arrest model—(1) in the positioning of the crack arrest transition curve relative to the crack initiation transition curve (Figure 5-4), and (2) in the scatter in the crack arrest toughness data about the mean transition curve (Figure 5-5 and Figure 5-6). Figure 5-8 shows these submodels overlaid on the flow diagram for the crack arrest model. The following paragraphs discuss the uncertainty treatment for each submodel.

The scatter on Figure 5-4 represents the statistical error in determining both T_o and T_{KIa}, which scales in proportion to $1/\sqrt{n}$ where n is the number of fracture toughness specimens tested (see ASTM E1921). This information suggests that the uncertainty in ΔRT_{ARREST} on Figure 5-4 is reducible, making it epistemic in nature. Therefore, FAVOR selects individual values of ΔRT_{ARREST} from the distribution shown in Figure 5-4 for each simulation run at the RT_{LB} value of interest.

From the physical model of cleavage crack arrest toughness described in Section 4.2.1, one concludes that the occurrence or nonoccurrence of crack arrest depends upon the interaction of a rapidly evolving stress state in front of a running crack with the distribution of defects in the material that inhibit dislocation motion. The barriers to dislocation motion include vacancy clusters, interstitial clusters, coherent and semicoherent particles, and other dislocations. These barriers are all of nanometer size and have interdefect spacings on the same size scale. Scatter in K_{Ia} data therefore occurs as a consequence of the randomness in the distribution of barriers to dislocation motion throughout the material. Because the distribution of these barriers throughout the lattice is random, and the conditions for arrest must be satisfied over a significant portion of the advancing crack front for arrest to occur, the uncertainty in K_{Ia} data is irreducible. For this reason the uncertainty in K_{Ia} is treated as aleatory. FAVOR achieves an aleatory representation of K_{Ia} uncertainty by executing the through-wall cracking model 100 times for each time step at which the crack initiation model estimates a nonzero conditional probability of crack initiation (CPI). Each of these 100 trials is a single, deterministic crack-arrest calculation, and each calculation is performed at a K_{la} toughness that is a different, randomly selected percentile of the uncertainty distributions illustrated in Figure 5-5 and Figure 5-6. The fraction of these 100 trials for which through-wall cracking is predicted provides an estimate of the percentage of CPI that is manifested in the conditional probability of vessel failure (CPF).

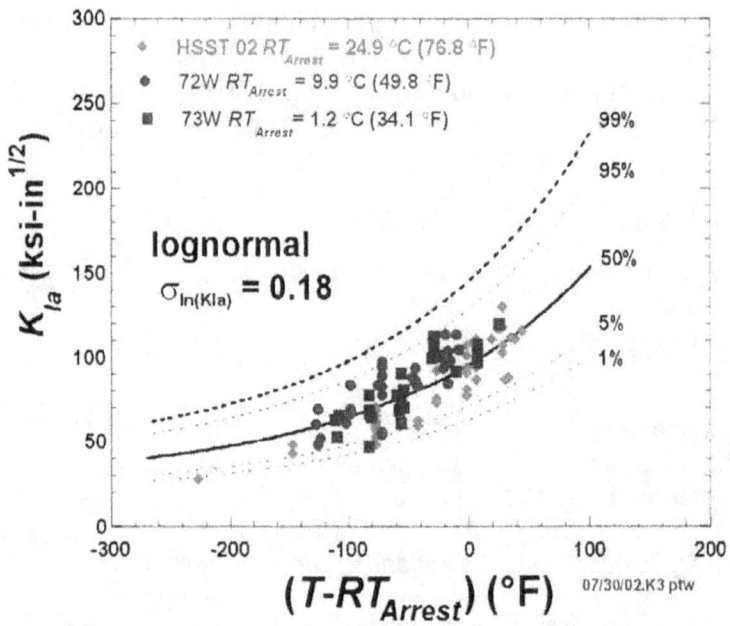

Figure 5-5 FAVOR K_{la} model based only on ASTM E1221 data

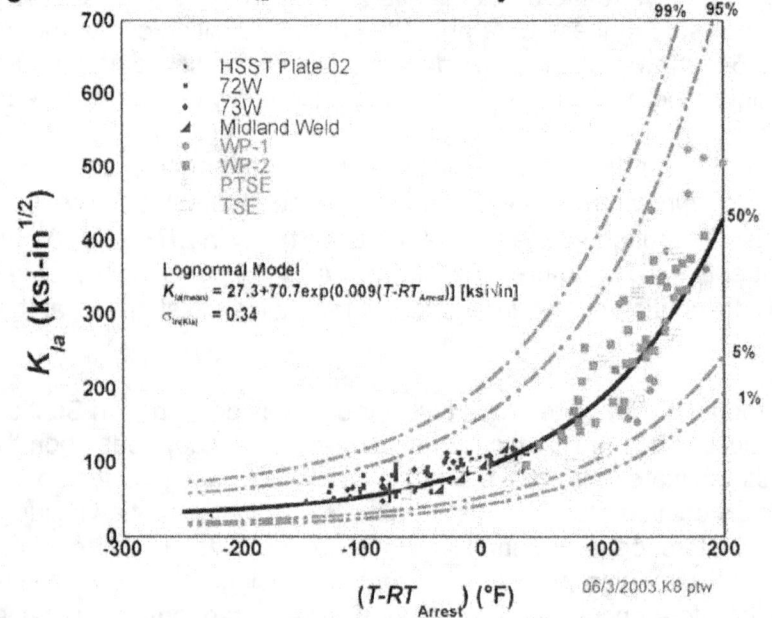

Figure 5-6 FAVOR K_{la} model based on ASTM E1221 data, wide-plate data, thermal shock experiment data, and PTS experiment data

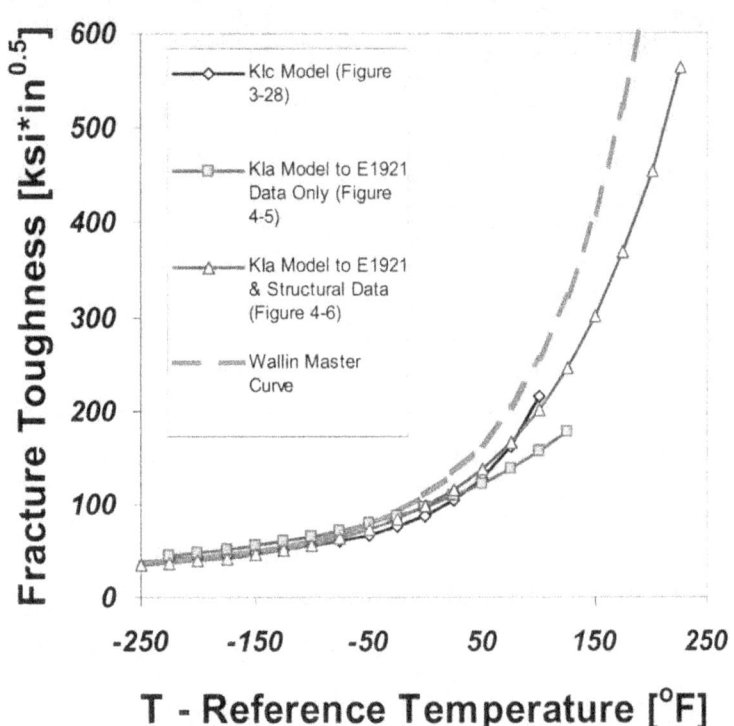

Figure 5-7 Comparison of the temperature dependencies of the FAVOR K_{Ic} model, the two FAVOR K_{Ia} models, and the Wallin Master Curve. Note that only median and mean curves are shown, and that the curves are only plotted over their calibrated range.

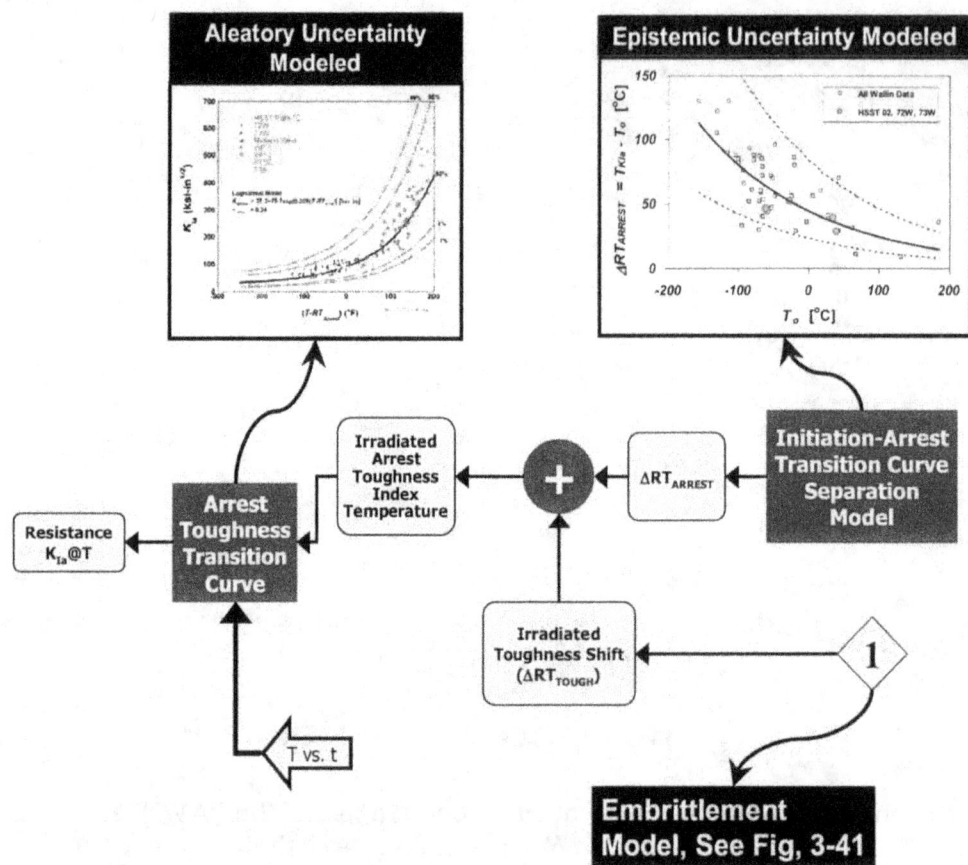

Figure 5-8 Diagrammatic summary of the FAVOR crack arrest model showing the recommended models, input values, and uncertainty treatment

5.2 Upper-Shelf Fracture Toughness Model

5.2.1 Need for an Upper-Shelf Model

No versions of FAVOR (and its predecessor codes OCA-P and VISA) earlier than FAVOR Version 03.1 include a model for upper-shelf (ductile) fracture toughness (see Williams 04). These earlier probabilistic codes assume that both K_{Ic} and K_{Ia} in the transition region could attain a maximum value of 200 ksi√in. This 200 ksi√in limit was rationalized on the basis that the American Society of Mechanical Engineers (ASME) K_{Ic} and K_{Ia} curves are capped at this value. While this is true, the 200 ksi√in limit, and the exclusion of ductile tearing as a potential consequence of pressurized thermal shock (PTS) loading, is at variance with physical reality for the following reasons:

- As illustrated previously (see Figure 5-6), crack arrest can occur at values considerably in excess of 200 ksi√in.

- By excluding the possibility of ductile tearing from the probabilistic model, one assumes that once the crack arrests, reinitiation by ductile tearing is not possible. However, a collection of J_{Ic} data for a wide variety of RPV steels (both low-upper shelf and nonlow-upper shelf, including irradiated and unirradiated; see Figure 5-9) demonstrate that 200 ksi√in, if anything, exceeds the resistance of most RPV steels to the initiation of ductile tearing over the temperature range of interest for PTS loading. Thus, if a crack arrested at a $K_{APPLIED}$ value of 200 ksi√in or above, some amount of ductile tearing would almost certainly follow.

For these reasons, ignoring the possibility of ductile crack initiation and tearing is both physically unrealistic and, potentially, nonconservative. Consequently, the NRC has incorporated a model of ductile crack initiation and tearing into this version of FAVOR. The following sections describe this model.

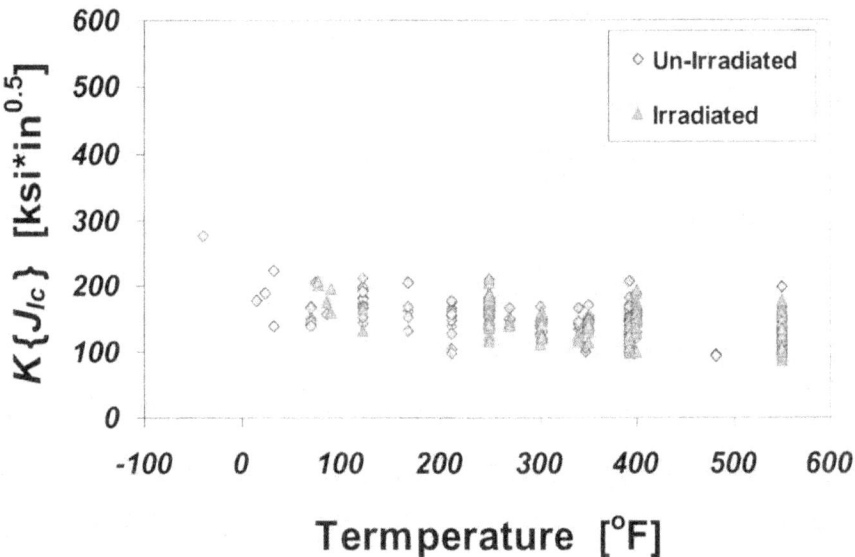

Figure 5-9 J_{Ic} **data for U.S. RPV steels (converted to *K*-units) before and after irradiation**
5.2.2   Characterization of Upper-Shelf Fracture Toughness Properties

In ferritic steels crack initiation occurs when there is sufficient stress and strain to initiate and grow microvoids in the material to a macroscopic size. Once crack initiation occurs, the crack extends as these microvoids continue to initiate and grow in the high stress/strain fields ahead of the crack tip and, eventually, coalesce into the main crack. This ductile crack growth behavior is characterized using a *J-R* curve determined according to the American Society for Testing and Materials (ASTM) Standard E-1820, or other similar procedures. Figure 5-10 illustrates schematically the results of a *J-R* curve test (the open and filled circles) and shows how the initiation fracture toughness (J_{Ic}) and the ductile tearing resistance (T_{mat} and *n*) are defined from these experimental data. These characterizations are described in greater detail below:

- Initiation Fracture Toughness (J_{Ic}). The current version of ASTM E1820 defines J_{Ic} as the value of *J* at the intersection of the 0.2 mm offset line (offset parallel to the blunting line) and a fit to the experimental *J-Δa* data of the form $J=C\Delta a^{n}$. Only *J-Δa* pairs between the 0.15 and 1.5 mm exclusion lines are used to develop this fit.

- Tearing Resistance (T_{mat} and *n*). The resistance to ductile tearing after crack initiation has been characterized in two (closely related) ways over the years, using T_{mat} (which is the normalized linear slope of all the *J-Δa* data between the 0.15 and 1.5 mm exclusion lines), and using *n*, which is the exponent of the function used in the determination of J_{Ic}.

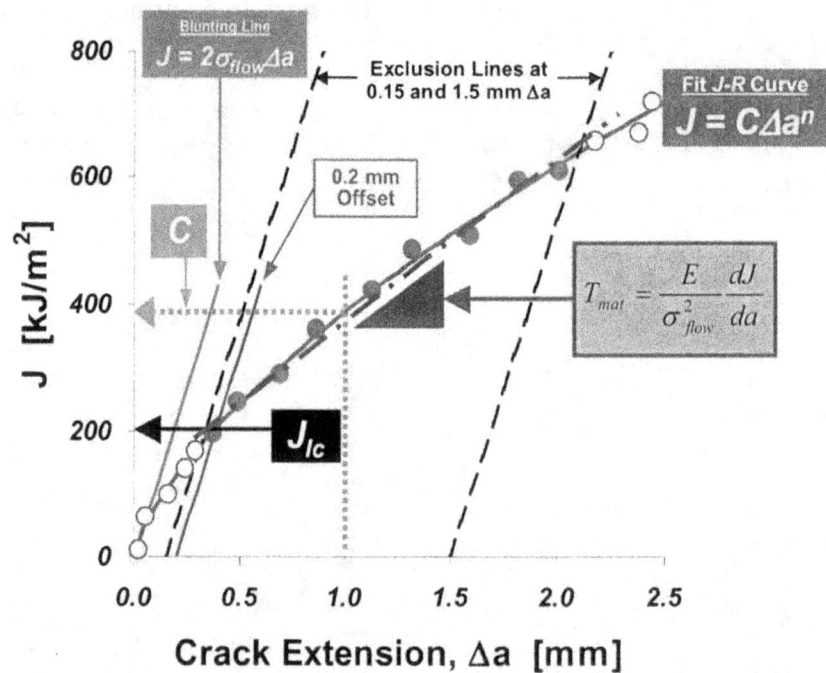

Figure 5-10 *J-Δa* data (filled and open circles) and fits and constructions prescribed by ASTM Standard E1820

5.2.3 Strategies for Estimating Upper-Shelf Fracture Toughness Properties

To assess the possibility of ductile crack initiation and tearing in FAVOR, a methodology is needed to estimate the upper-shelf fracture toughness properties illustrated in Figure 5-10, the uncertainty in these properties, and how they change with both temperature and irradiation. The NRC Reactor Vessel Integrity Database (RVID) includes measurements of the energy needed to break Charpy V-notch (CVN) specimens on the upper shelf, a quantity called USE (RVID2]. Indeed, most strategies for estimating the resistance to crack initiation (J_{lc}) and to further stable tearing (*J-R*) on the upper shelf are based on correlations with USE (see Eason 91 and Wallin 01 as examples). However, relationships between USE and the real upper-shelf fracture toughness properties illustrated in Figure 5-10 exhibit notoriously poor correlation coefficients (see Figure 5-11). This lack of correlation suggests that different physical mechanisms underlie energy absorption in the CVN test versus the resistance to ductile crack initiation and propagation from a preexisting defect. Additionally, these low correlation coefficients engender little confidence in the reliability of calculations made on the basis of such relationships.

Recently EricksonKirk proposed a new strategy for estimating upper-shelf fracture toughness properties (see EricksonKirk 04b). This new model does not rely on Charpy correlations in any way, and features an explicit treatment of the uncertainty in upper-shelf toughness data. Additionally, the new model estimates the upper-shelf toughness properties from the cleavage fracture toughness transition temperature (T_o), a relationship motivated both by trends in fracture toughness data and by physical considerations.

The following section describes this new model, discusses its basis, and then details its implementation in FAVOR (see Section 5.2.5).

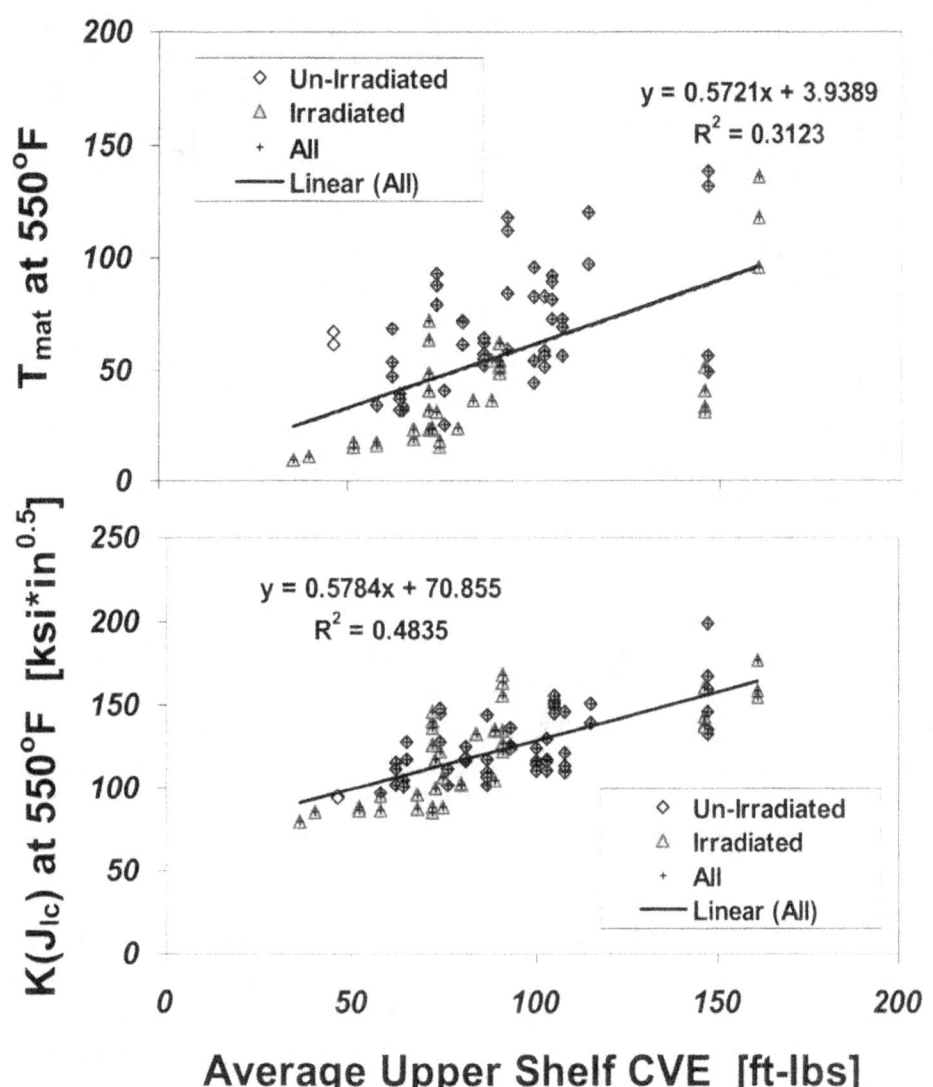

Figure 5-11 Relationship between upper-shelf toughness measures and CVN USE for irradiated and unirradiated RPV steels

5.2.4 New Upper-Shelf Model Proposed by EricksonKirk

In work recently completed under Electric Power Research Institute (EPRI) sponsorship, EricksonKirk proposed a model of upper-shelf toughness that does not rely on correlations with CVN properties (see EricksonKirk 04b). The following subsections describe both the empirical and physical bases for this model. This information, which is largely copied directly from EricksonKirk 04b, has been included with EPRI's permission.

5.2.4.1   Empirical Basis

5.2.4.1.1   Database

The data in Figure 5-12 demonstrate that RPV steels often viewed as being different (i.e., plate *vs.* weld, irradiated *vs.* unirradiated, or low upper-shelf materials) exhibit a strikingly similar temperature dependency on the upper shelf, suggesting the possibility for developing a general model of fracture toughness behavior on the upper shelf. To investigate this possibility, EricksonKirk compiled a database of nearly 400 ductile fracture toughness data records for nuclear RPV steels that show the effects of both temperature and radiation damage from the literature sources listed in Table 5-1. Additionally, the literature sources listed in

Table 5-2 provide 85 ductile fracture toughness data records for other ferritic steels, as well as for a tempered martensitic steel. In addition to fracture toughness data, these sources also provide information concerning chemical composition and product form (see Table 5-3), strength, and CVN USE (see Table 5-4). This database includes ferritic RPV steels having a wide range of CVN USE inclusive of both low upper-shelf materials, as well as the high toughness, low-copper plate materials characteristic of modern steelmaking practice. Examination of the information in Table 5-3 and Table 5-4 reveals that this database covers a wide range of both chemical composition (copper from 0.04 to 0.42 weight percent; nickel from 0.1 to 0.7 weight percent) and radiation exposure (fluences up to 2.06×10^{19} n/cm^2).

5.2.4.1.2 Temperature Dependency of J_{Ic}

The common upper-shelf temperature dependency of J_{Ic} exhibited by different steels (see Figure 5-12) motivates an examination of the temperature dependency of J_{Ic} for all of the steels in the database (see Figure 5-13). While these J_{Ic} values exhibit some trends with temperature, it is also clear that different steels can have considerably different toughness levels at the same temperature. To focus attention strictly on the temperature dependency of J_{Ic} the material-condition-specific toughness level is removed from the data as follows:

- Calculate the average of the measured values of J_{Ic} at 288 °C (550 °F) for each material condition (a unique combination of material composition and radiation exposure defines a material condition).

- Calculate the difference between average J_{Ic} at 288 °C (550 °F) (J_{Ic288}) and each measured value of J_{Ic} in the database.

The use of 288 °C (550 °F) as the normalization temperature is completely arbitrary. The value of 288 °C (550 °F) was selected strictly because J_{Ic} measurements are most commonly reported at this temperature for nuclear RPV steels (288 °C (550 °F) is the nominal operating temperature of a pressurized-water reactor (PWR)). Thus, normalization at 288 °C (550 °F) admitted the largest quantity of data for further analysis. Figure 5-14 demonstrates that this normalization removes a considerable degree of the scatter from the data that results from material-condition-specific toughness levels. Furthermore, Figure 5-14 reveals a temperature dependency of J_{Ic} that is the same for a wide range of material conditions. Specifically, all of the ferritic steels exhibit the same temperature dependency irrespective of product form, chemical composition, irradiation level, strength level, or strengthening mechanism. Only HY-80 (a tempered martensitic steel) fails to exhibit this common temperature dependency, suggesting that the upper-shelf temperature dependence is only influenced by the crystal structure of the material.

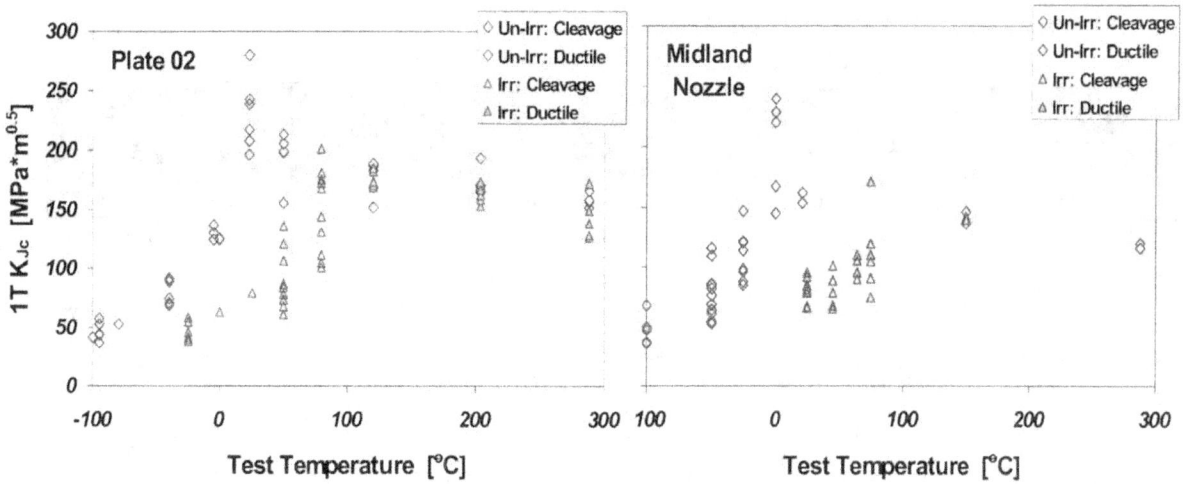

Figure 5-12 Transition and upper-shelf fracture toughness data for an A533B Steel (HSST Plate 02, left) and for a Linde 80 weld (Midland nozzle course weld, right) indicating that RPV steels normally viewed as being different exhibit strikingly similar temperature dependency on the upper shelf

Table 5-1 Sources of Upper-Shelf Data for RPV Steels

Materials	Reference
Seven welds designated 61W through 67W	70
Midland beltline and nozzle course welds	71
Plate 02 and four welds designated 68W through 71W	72
Two welds designated W8A and W9A	73
Characterization of the PTSE-2 plate material	74
JAERI Plates A & B	75
Note. References in this table are given in Appendix E.	

Table 5-2 Sources of Upper-Shelf Data for Other Ferritic and Tempered Martensitic Steels

Materials	Reference
One ASTM-A710 plate and one HY-80 plate	76
Two ASTM-A710 plates and eight HSLA-100 plates	77
Two ASTM-A710 plates	78
One ASTM-A710 plate	79
Note. References in this table are given in Appendix E.	

Table 5-3 Chemical Composition Information

Ref.	HSST ID	Flux Lot	C	Mn	P	S	Si	Cr	Ni	Mo	Cu	V
72	Plate 02	N/A	0.23	1.55	0.009	0.014	0.2	0.04	0.67	0.53	0.14	0.003
71	Midland Beltline	Linde 80	0.083	1.607	0.017	0.006	0.622	0.1	0.574	0.41	0.256	0.006
71	Midland Nozzle	Linde 80	0.083	1.604	0.016	0.007	0.605	0.11	0.574	0.39	0.29	0.008
73	W8A	Linde 80	0.083	1.33	0.011	0.016	0.77	0.12	0.59	0.47	0.39	0.003
73	W9A	Linde 0091	0.19	1.24	0.01	0.008	0.23	0.1	0.7	0.49	0.39	
72	68W	Linde 0091	0.15	1.38	0.008	0.009	0.16	0.04	0.13	0.6	0.04	0.007
72	69W	Linde 0091	0.14	1.19	0.01	0.009	0.19	0.09	0.1	0.54	0.12	0.005
72	70W	Linde 0124	0.1	1.48	0.011	0.011	0.44	0.13	0.63	0.47	0.056	0.004
72	71W	Linde 80	0.12	1.58	0.011	0.011	0.54	0.12	0.63	0.45	0.046	0.005
70	61W	Linde 80 btwn A533B	0.09	1.48	0.02	0.014	0.57	0.16	0.63	0.37	0.28	0.005
70	62W	Linde 80 btwn A508	0.083	1.51	0.16	0.007	0.59	0.12	0.537	0.377	0.21	0.01
70	63W	Linde 80 btwn A508	0.098	1.65	0.016	0.011	0.63	0.095	0.685	0.427	0.299	0.011
70	64W	Linde 80 btwn A508	0.085	1.59	0.014	0.015	0.52	0.092	0.66	0.42	0.35	0.007
70	65W	Linde 80 btwn A508	0.08	1.45	0.015	0.015	0.48	0.088	0.597	0.385	0.215	0.006
70	66W	Linde 80 btwn A508	0.092	1.63	0.018	0.009	0.54	0.105	0.595	0.4	0.42	0.009
70	67W	Linde 80 btwn A508	0.082	1.44	0.011	0.012	0.5	0.089	0.59	0.39	0.265	0.007
74	PTSE2 Post	N/A	0.13	0.40	0.009	0.018	0.19	2.25	0.11	0.94	0.08	
75	Onizawa A – 70s Japanese Plate	N/A	0.19	1.30	0.015	0.010	0.30	0.17	0.68	0.53	0.16	
75	Onizawa A – Late 80s Japanese Plate	N/A	0.19	1.43	0.004	0.001	0.19	0.13	0.65	0.50	0.04	

Note. References in this table are given in Appendix E.

Table 5-4 Strength and Upper-Shelf Energy Information

Product Form	Citation	HSST Designation	Flux Lot	Neutron Fluence / 10^{19} [n/cm^2]	Room Temp. σ_{ys} [MPa]	Room Temp. σ_{uts} [MPa]	Upper-Shelf CVN Energy [J]
Plate	72	Plate 02		0	467	622	142
	72	Plate 02		1.77	613	751	123
	74	PTSE2 Post		0	469	621	62
	75	Onizawa A		0	469	614	150
	75	Onizawa B		0	462	600	207
Weld	71	Midland Beltline	Linde 80	0	407	586	88
	71	Midland Beltline	Linde 80	1	646	747	80
	71	Midland Nozzle	Linde 80	0	545	655	87
	71	Midland Nozzle	Linde 80	1	701	792	68
	73	W8A	Linde 80	0	481	604	79
	73	W8A	Linde 80	1.57	658	751	54
	73	W8A	Linde 80	2.065	668	749	49
	73	W9A	Linde 0091	0	564	659	156
	73	W9A	Linde 0091	1.42	709	785	114
	73	W9A	Linde 0091	2.065	726	799	100
	72	Weld 68W	Linde 0091	0	553	641	199
	72	Weld 68W	Linde 0091	1.35	564	645	218
	72	Weld 69W	Linde 0091	0	638	722	199
	72	Weld 69W	Linde 0091	1.22	711	784	198
	72	Weld 70W	Linde 0124	0	478	594	100
	72	Weld 70W	Linde 0124	1.65	534	649	98
	72	Weld 71W	Linde 80	0	469	599	110
	72	Weld 71W	Linde 80	1.65	539	649	121
	70	61W		0	480	626	84
	70	61W		1.24	600	722	70
	70	62W		0	473	590	93
	70	62W		1.36	615	702	80
	70	63W		0	488	600	87
	70	63W		1.19	625	710	68
	70	64W		0	469	600	100
	70	64W		0.68	579	696	75
	70	65W		0	448	572	108
	70	65W		0.7	572	662	72
	70	66W		0	531	655	76
	70	66W		0.94	641	745	58
	70	67W		0	462	607	103
	70	67W		0.89	579	690	73

Note. References in this table are given in Appendix E.

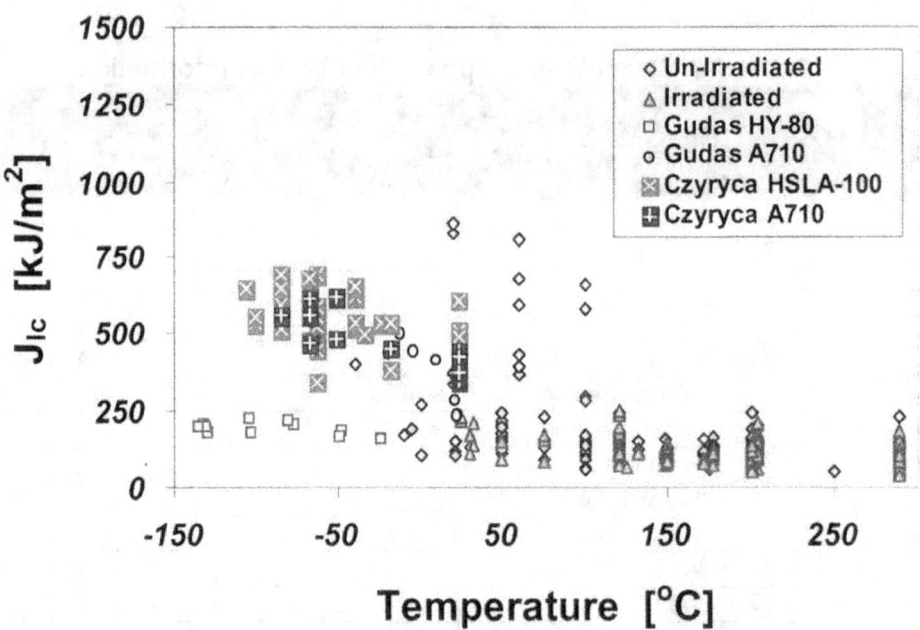

Figure 5-13 Relationship between J_{Ic} and temperature for irradiated and unirradiated RPV steels

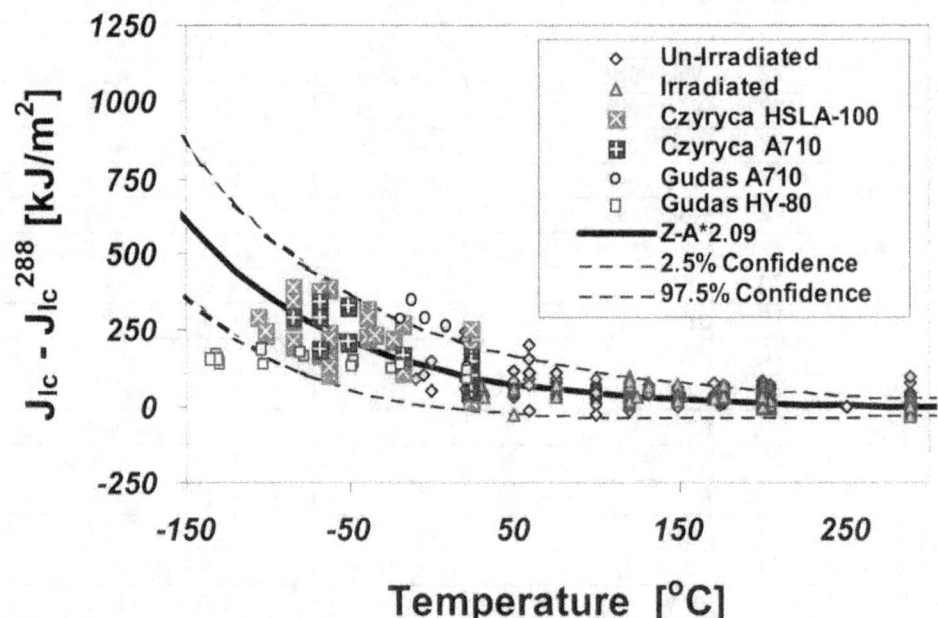

Figure 5-14 Relationship between J_{Ic} and temperature for irradiated and unirradiated RPV steels. Here the J_{Ic} values have been normalized relative to the average value of J_{Ic} at 288 °C (550 °F).

The fit to the normalized J_{Ic} data on Figure 5-14 is as follows:

Eq. 5-7

$$J_{Ic}(T) - J_{Ic}(288°C) = 2.09\{C_1 \cdot \exp[\tau] - \sigma_{ref}\}$$
$$\tau = [-C_2(T + 273.15) + C_3(T + 273.15) \cdot \ln(\dot{\varepsilon})]$$

where T_{ref} = 288 °C (or 561K), C_1 = 1033 MPa, C_2 = 0.00698/K, C_3 = 0.000415/K, $\dot{\varepsilon}$ = 0.0004/s, and σ_{ref} = 3.33 MPa. This fit is of the same functional form as the dislocation-based model for flow behavior adopted by Zerilli and Armstrong (see Zerilli 87). This correspondence suggests a physical basis for the consistent trend; Section 5.2.4.2 will discuss this in more detail.

5.2.4.1.3 Relationship Between Fracture Toughness in the Transition and Fracture Toughness on the Upper Shelf

Identifying this common temperature dependency permits the use of the J_{Ic} database together with the Master Curve for transition fracture toughness proposed by Wallin (see Eq. 4-3) to identify a temperature above which upper-shelf behavior would be expected. As illustrated in Figure 5-15, fitting a Master Curve through K_{Jc} data and fitting Eq. 5-7 through J_{Ic} data defines the temperature at which the curves intersect, which is labeled T_{US}. Like T_o, T_{US} is defined in a somewhat arbitrary fashion. T_{US} carries with it no particular physical interpretation (it is not, for example, the temperature above which cleavage fracture cannot occur). However, defining T_{US} using identified temperature dependencies in transition and on the upper shelf that are common to a broad range of steels offers the advantage of reduced ambiguity and, therefore, improved accuracy in the T_{US} index temperature definition.

Figure 5-16 shows the variation of T_{US} with T_o for the steels listed in Table 5-5. The data on this plot include both irradiated and unirradiated nuclear RPV steels (including plates and welds made using several flux types, including so-called "low upper-shelf" welds), as well as the higher-strength, high toughness, copper-precipitation-hardened steels ASTM A710 and HSLA-100 used in surface ship fabrication. The data on Figure 5-16 exhibit a consistent trend over T_o values spanning nearly 300 °C (540 °F). Dr. Kim Wallin of VTT in Finland has corroborated this trend and provided the additional data shown in Figure 5-17. Dr. Wallin's data add more materials to the database (VVER (in Russian, vodo-vodianyj energeticheskij reactor, or water moderated, water cooled power reactor) steels and a ferritic stainless steel) and expand the range of T_o values covered by the T_{US} relationship to nearly 400 °C (720 °F). The fit to the complete database (both that reported in EricksonKirk 04b and that provided by Wallin) is as follows:

Eq. 5-8 $T_{US} = 50.1 + 0.794T_o$, T_o and T_{US} in °C

The consistency of this trend across such a wide range of ferritic steels strongly suggests an underlying physical basis for the relationship between upper-shelf behavior and transition behavior (i.e., materials exhibiting higher transition temperatures behavior will also exhibit lower upper-shelf fracture toughness). Section 5.2.4.2 suggests a physical basis for this trend (see EricksonKirk 04b), while Section 5.2.5 utilizes this trend to develop a model for FAVOR which couples transition fracture toughness and upper-shelf fracture toughness using Eq. 5-7.

5.2.4.2   Physical Basis

Limited information has been written regarding a relationship between transition and upper-shelf toughness behavior. EricksonKirk 04b proposes the following explanation of a physical basis for the observed trends in the data.

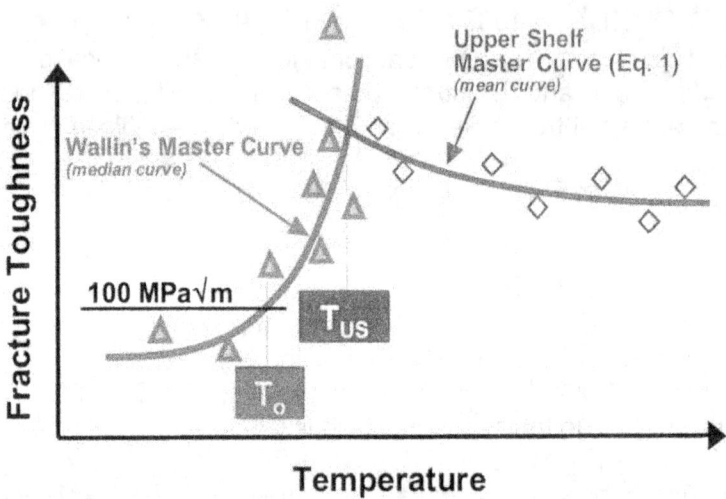

Figure 5-15 Schematic illustrating the relationship between transition and upper-shelf toughness and defining the value T_{US} as the intersection of the Wallin Master Curve and the upper-shelf Master Curve

The equation used to describe the temperature dependence of the upper-shelf fracture toughness behavior, Eq. 5-7, follows the same temperature dependence as that shown by Zerilli and Armstrong (ZA) to describe the flow properties of ferritic steels (see Zerrili 87). This is wholly expected since the upper-shelf behavior is defined by the temperature dependence of dislocation movement through the lattice. As temperature increases, lattice atom vibration amplitude increases, reducing the barriers to dislocation motion presented by the lattice atoms. As with transition toughness, this translates to a toughness temperature dependence that is closely related to that proposed by ZA for the flow stress. But while the transition toughness increases with increasing dislocation mobility, the upper-shelf toughness decreases because dislocation structures remain more homogeneous, allowing large amounts of plastic strain with little hardening at lower energies.

The fact that both transition and upper-shelf toughness behavior can be predicted from the same parameter (T_o) again lies within an understanding of the relationship of toughness to flow stress. Both are a measure of resistance of a material to crack extension or to dislocation motion. Furthermore, they are inversely related by the fact that cleavage fracture occurs when dislocations stop moving, while ductile fracture occurs when dislocations continue to move through the matrix to grow holes in the material. Therefore, the cleavage fracture toughness is simply a measure of the cessation of ductile behavior in the material. Transition behavior measures the energy absorbed by dislocation motion prior to accumulating at a microstructural nonhomogeneity, while upper-shelf behavior measures the energy absorbed when dislocation motion does not stop, but continues to final material separation. Because cleavage cracking stops at temperatures above which ductile hole growth begins, the two parameters should be closely related, and they should exhibit an inverse temperature dependence with each other.

The temperature dependence of the two curves is thus expected to be based on the temperature dependence of dislocation motion through the lattice atoms, which, as shown by ZA, remains the same throughout the temperature region in which dislocations move. Indeed, both curves have been shown to exhibit a ZA-predicted temperature relationship (see Natishan 01, EricksonKirk 04b). The only metallurgical parameters that will affect the temperature dependence of dislocation motion within BCC materials is the lattice atom structure itself, as lattice atom vibration is the only metallurgical parameter affected by a change in thermal energy (within the ranges of fracture toughness testing; see Zerilli 87). Because both K_{Ic} and J_{Ic} are measures of the energy dissipated by dislocation motion through the matrix material prior to fracture, and because the lattice structure is the same (BCC with the same lattice parameter) for all ferritic steels, then the temperature dependence of fracture toughness behavior is expected to be closely related. This relationship should only break down for materials with a differing lattice structure (e.g., martensitic steels).

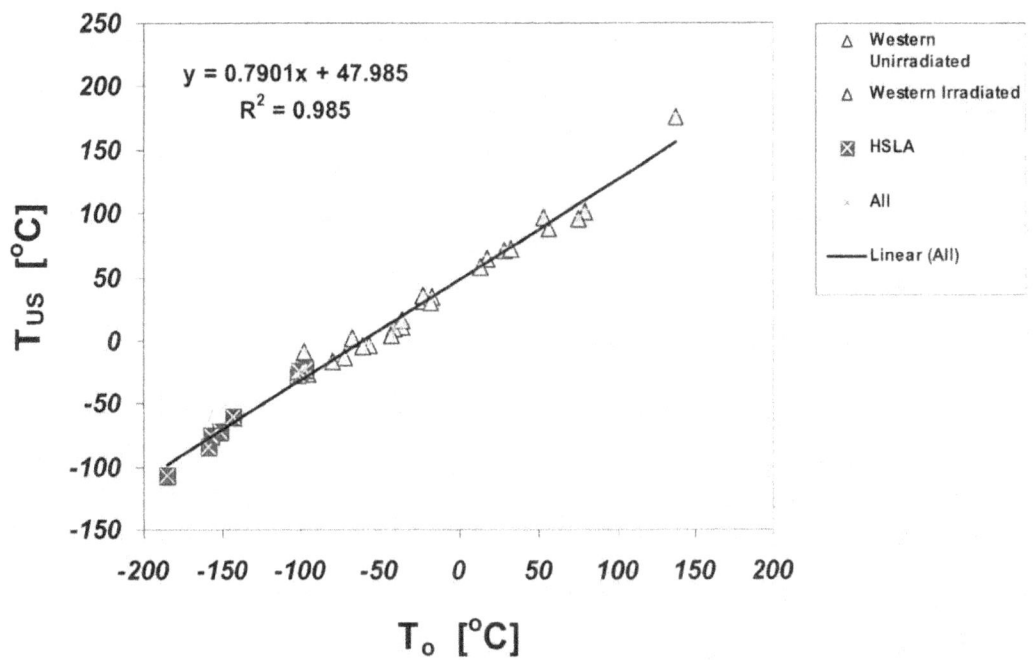

Figure 5-16 Relationship between T_{US} and T_o reported in EricksonKirk 04b

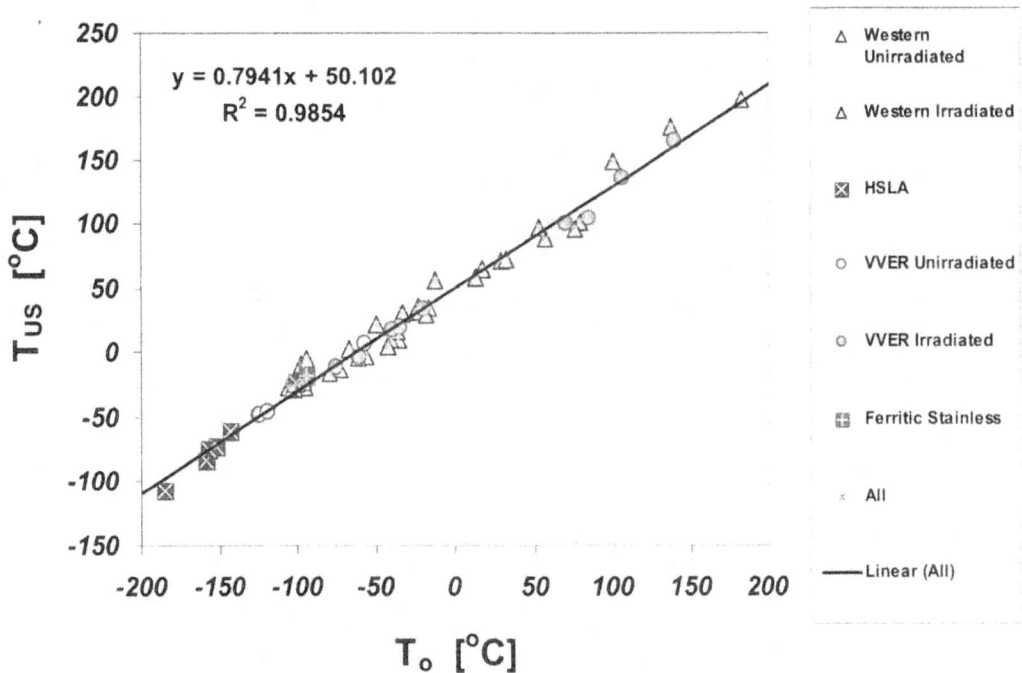

Figure 5-17 Data from Figure 5-16 augmented by data provided by Wallin

Table 5-5 Summary of Data Used in EricksonKirk 04b to Establish the Relationship between T_o and T_{US}

Citation	Data Set ID	Product Form	Fluence [n/cm²]	T_o [°C]	$T_{Upper Shelf}$ [°C]
71	Midland Belt	Linde 80	0	-56	13
71	Midland Nozzle	Linde 80	0	-36	26
72	Plate 02	A533B	0	-25	46
72	68W	Linde 0091	0	-95	-10
72	69W	Linde 0091	0	-17	51
72	70W	Linde 0124	0	-61	13
72	71W	Linde 80	0	-41	26
80	ks-01	Linde 80	0	-23	41
73	W8A	Linde 80	0	-42	20
73	W9A	Linde 0091	0	-72	3
75	JAERI Plate A	Like A533B	0	-67	8
75	JAERI Plate B	Like A533B	0	-98	-3
76	A710 - GGO	Plate		-98	-20
78	HSLA-100, GLG-1	Plate		-152	-68
78	HSLA-100, GLF-1.25	Plate		-143	-56
78	HSLA-100, GKO-2	Plate		-102	-25
78	A710 - GAW	Plate		-159	-82
79	A710 - GFF	Plate		-185	-103
79	A710 - GGN	Plate		-97	-20
78	HSLA-100, GLC-1	Plate		-157	-70
78	HSLA-100, GKN-1.25	Plate		-101	-21
71	Midland Belt	Linde 80	1E+19	29	76
71	Midland Nozzle	Linde 80	1E+19	57	93
72	Plate 02	A533B	1.77E+19	53	113
72	68W	Linde 0091	1.35E+19	-80	0
72	69W	Linde 0091	1.22E+19	17	81
72	70W	Linde 0124	1.65E+19	-36	33
72	71W	Linde 80	1.65E+19	-18	47
80	ks-01	Linde 80	8.00E+18	137	182
73	W8A	Linde 80	1.5E+19	80	119
73	W8A	Linde 80	2.1E+19	76	114
73	W9A	Linde 0091	1.5E+19	13	75
73	W9A	Linde 0091	2.1E+19	32	89

5.2.5   Incorporation of the New Upper-Shelf Model into FAVOR

The model developed in EricksonKirk 04b provides both a temperature dependency for upper-shelf fracture toughness and a means to locate the upper shelf relative to the transition fracture index temperature, T_o. To provide a complete model of fracture toughness on the upper shelf for use in FAVOR, the following four additional items are needed:

(1)   a mathematical description of how upper-shelf fracture toughness is estimated only from T_o (this merely involves algebraic manipulation of equations already presented and is included for completeness)

(2)   a discussion of the origin of the uncertainty in the T_{US} vs. T_o model (Eq. 5-8)

(3)   a description of the scatter in J_{Ic} on the upper shelf

(4)   a characterization of the J-R curve to account for stable tearing

The following four sections present these items in order.

5.2.5.1 Procedure to Establish the Temperature Variation of Fracture Toughness for Both Transition and Upper Shelf Using Only T_o

A model of ferritic steel toughness that accounts for fracture mode transition behavior, upper-shelf behavior, and the interaction between these two different fracture modes can be constructed based on Wallin's Master Curve (Eq. 4-3), the T_{US} vs. T_o relationship (Eq. 5-8), and the upper-shelf Master Curve (Eq. 5-7). Using these equations it is possible, as described in the steps below, to estimate the complete variation of initiation fracture toughness with temperature in both the transition regime and on the upper shelf based only on a measurement of T_o. This description repeats equations used elsewhere in the interest of clarity.

(1)   Estimate T_o using ASTM E1921.

(2)   Convert Wallin's Master Curve to J units:

$$J_c = \frac{\{30 + 70 \cdot \exp(0.019(T - T_o))\}^2 (1 - v^2)}{E}$$

where $E = \{207200 - 57.1 \cdot T\}$ T is in °C, E is in MPa, and $v = 0.3$.

(3)   Calculate T_{US} using the T_o value from step 1.

$T_{US} = 50.1 + 0.794 T_o$, (temperature in °C)

(4)   Calculate J_c using the equation in step 2 at T_{US} from step 3. Call this value $J_{c(US)}$.

(5)   Calculate ΔJ_{Ic} at T_{US} using the following equation. Call this value $\Delta J_{Ic(US)}$.

$$\Delta J_{Ic} \equiv J_{Ic} - J_{Ic}^{288} = 2.09\{C_1 \cdot \exp[\tau] - \sigma_{ref}\}, \quad \tau = [-C_2(T + 273.15) + C_3(T + 273.15) \cdot \ln(\dot{\varepsilon})]$$

where
    T_{ref} = 288 °C (or 561K),
    C_1 = 1033 MPa,
    C_2 = 0.00698/K,
    C_3 = 0.000415/K,
    $\dot{\varepsilon}$ = 0.0004/s, and
    σ_{ref} = 3.33 MPa.

(6)   Let $J_{ADJUST} = \{J_{c(US)}\text{-}\Delta J_{Ic(US)}\}$ using $J_{c(US)}$ from step 4 and $\Delta J_{Ic(US)}$ from step 5.

(7)   The variation of the mean value of J_{Ic} with temperature can now be estimated as follows:

$$J_{Ic} = 2.09\{C_1 \cdot \exp[\tau] - \sigma_{ref}\} + J_{ADJUST}, \quad \tau = [-C_2(T + 273.15) + C_3(T + 273.15) \cdot \ln(\dot{\varepsilon})]$$

The equivalence of the median value of J_c and the mean value of J_{Ic} at T_{US} is enforced by the value of J_{ADJUST} calculated in step 6. Conversion of this equation to K-units can be made by using the conventional relationship between J and K in plane strain:

$$K_{Jc} = \sqrt{\frac{J \cdot E}{1 - v^2}}$$

The FAVOR theory manual discusses how this procedure is adapted to be compatible with the initiation fracture toughness index temperature used in FAVOR Version 04.1 (RT_{LB}) (see Williams 04).

5.2.5.2   Uncertainty in the T_o vs. T_{US} Relationship

The standard deviation of the fit represented by Eq. 5-8 is 8 °C (15 °F). It is possible to assess the origin of this 8 °C (15 °F) uncertainty by examining how the residual between each (T_{US}, T_o) data point and the best fit of Eq. 5-8 depends (or not) on the number of K_{Jc} and J_{Ic} values used to estimate T_o and T_{US}, as follows:

- If the origin of the uncertainty about the best fit represented by Eq. 5-8 is error in accurately resolving the measured values of T_o and T_{US} because of limited experimental K_{Jc} and J_{Ic} data, then the residual between each (T_{US}, T_o) data point and the best fit of Eq. 5-8 will become smaller as the number of K_{Jc} and J_{Ic} values used to estimate T_o and T_{US} increases.

- If the origin of the uncertainty about the best fit represented by Eq. 5-8 arises because of material-dependent differences in the T_{US} vs. T_o relationship, then the residual between each (T_{US}, T_o) data point and the best fit of Eq. 5-8 will be independent of the number of K_{Jc} and J_{Ic} values used to estimate T_o and T_{US}.

The information in Figure 5-18 exhibits the trend expected from experimental error in accurately resolving the measured values of T_o and T_{US}, suggesting that the relationship between transition fracture toughness and upper-shelf fracture toughness shown in Figure 5-17, and quantified by Eq. 5-8, can be regarded as a best estimate, applicable to a very broad class of materials. Simulation in FAVOR of the uncertainty in Eq. 5-8 is therefore inappropriate because the relationship is not expected to vary from material to material.

5.2.5.3   Scatter in J_{Ic} Data

Establishing the scatter bands about Eq. 5-7, as shown in Figure 5-14, began by calculating the difference between each measured ΔJ_{Ic} value in the database and the ΔJ_{Ic} value predicted by the fit (Eq. 5-7). Figure 5-19 shows the variation of these ΔJ_{Ic} residuals with temperature. The residuals were then divided into various bins having finite temperature ranges, and the standard deviation of the residuals was calculated in each bin. Figure 5-19 also shows the variation of·

these standard deviations with temperature. The fit to the standard deviation values shown in Figure 5-19, together with Eq. 5-7, establish the scatter bands shown in Figure 5-14.

The FAVOR theory manual discusses the implementation of this information into FAVOR Version 04.1 (see Williams 04).

5.2.5.4   Estimation of the J-R Curve Tearing Parameter, m

To assess the ability of the RPV to sustain crack propagation by ductile tearing, an estimate of the J-R curve tearing parameter, m, is needed (see Figure 5-10). This parameter is not frequently reported in the literature. Indeed, of the literature data summarized in Table 5-1 through Table 5-5, only one citation (McGowan 88) reports this information. Figure 5-20 shows the McGowan data. The FAVOR theory manual discusses the incorporation of this information into FAVOR Version 4.1 (see Williams 04).

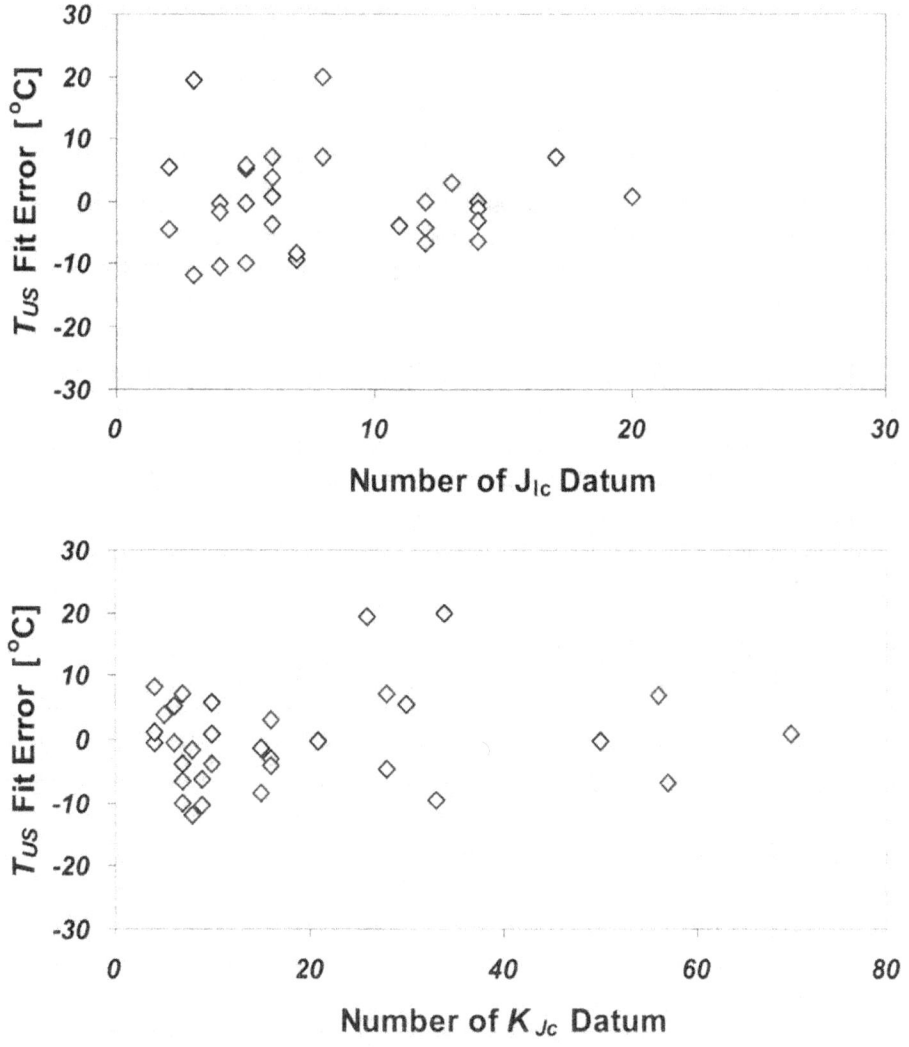

Figure 5-18 Effect of number of K_{Jc} and J_{Ic} data that underlie the T_o and T_{US} values on Figure 5-17 on how well an individual T_{US} value agrees with the fit (Eq. 5-8) to all of the data

5.2.6 Summary of Model and Uncertainty Treatment

Sections 5.2.4 and 5.2.5 established the various relationships necessary to build a probabilistic upper-shelf tearing model for FAVOR. This section summarizes this model, with reference to Figure 5-21.

Based on the value of the irradiated initiation fracture toughness index temperature, a value of T_{US} is estimated using Eq. 5-8. The uncertainty in the irradiated initiation fracture toughness index temperature was already simulated (see discussion of Figure 4-29 and Figure 4-42) and so does not need to be resimulated. The uncertainty Eq. 5-8 is small and, more importantly, results from the inability to accurately resolve the relationship using experimental data sets rather than from material-dependent differences in the underlying relationship (see discussion in Section 5.2.5.2). For these reasons, the uncertainty Eq. 5-8 is not simulated.

The value of T_{US} is used, along with the procedure detailed in Section 5.2.5.1, to estimate the variation of J_{lc} with temperature. Information in Sections 5.2.5.3 and 5.2.5.4 describes and quantifies the aleatory uncertainties in J_{lc} and in the J-R curve exponent m, respectively. FAVOR simulates these aleatory uncertainties.

5.3   Material Property Gradient Model

The material property gradient model includes two parts, (1) a relationship between crack initiation and crack arrest toughness (see Section 5.3.1), and (2) a relationship for welds that describes how the chemical composition varies through the thickness of the vessel wall (see Section 5.3.2).

5.3.1   Relationship between Crack Initiation and Crack Arrest Model

When performing a probabilistic fracture calculation, cracks may initiate, arrest, and reinitiate in the same simulation run. Consequently, a relationship is needed between K_{lc} and K_{la} data for the same material. Certainly the information in Figure 5-4 is needed because this specifies the temperature separation between the K_{lc} and K_{la} curves. However, mathematical simulation of crack initiation, arrest, and subsequent reinitiation also requires the following information:

- The arrest calculation requires the knowledge of whether permissible values of K_{la} are influenced in any way by the value of K_{lc} at which the crack initiated.

- The reinitiation calculation requires the knowledge of whether the permissible values of K_{lc} are influenced in any way by the value of K_{la} at which the crack arrested.

The following sections address these questions.

5.3.1.1   Influence of K_{lc} on Subsequent K_{la} Values

Crack arrest toughness is undefined at values above the crack initiation toughness because the fact that the crack has initiated means that arrest was not possible. Thus, the value of crack initiation toughness ($\hat{K}_{lc}$) establishes the maximum allowable K_{la} for temperature at which crack initiation occurred ($\hat{T}$). As the crack propagates deeper into the vessel wall, the temperature increases above $\hat{T}$ for PTS loadings, so restricting the maximum allowable K_{la} to $\hat{K}_{lc}$ is no longer appropriate. For temperatures in excess of $\hat{T}$, the maximum allowable K_{Ia} therefore cannot exceed the K_{Ia} value of the same percentile as the value $\hat{K}_{Ic}$ occupies in the K_{Ia} distribution at temperature $\hat{T}$. Figure 5-22 provides an illustration of these concepts.

Having established a physical rationale supporting adoption of the percentile corresponding to ($\hat{K}_{Ic}$, $\hat{T}$) as the maximum value of the K_{Ia} distribution, it is also necessary to specify how the distribution of K_{Ia} values in Eq. 4-7 below this limit is altered. The NRC considered the following two methods:

(1)   Truncate the K_{Ia} distribution established in Eq. 4-7 at the percentile corresponding to ($\hat{K}_{Ic}$, $\hat{T}$), but make no other changes to the distribution.

(2)   Scale the K_{Ia} distribution established in Eq. 4-7 so that some high percentile value (in the compressed distribution) corresponds to the percentile at ($\hat{K}_{Ic}$, $\hat{T}$) (in the unscaled distribution).

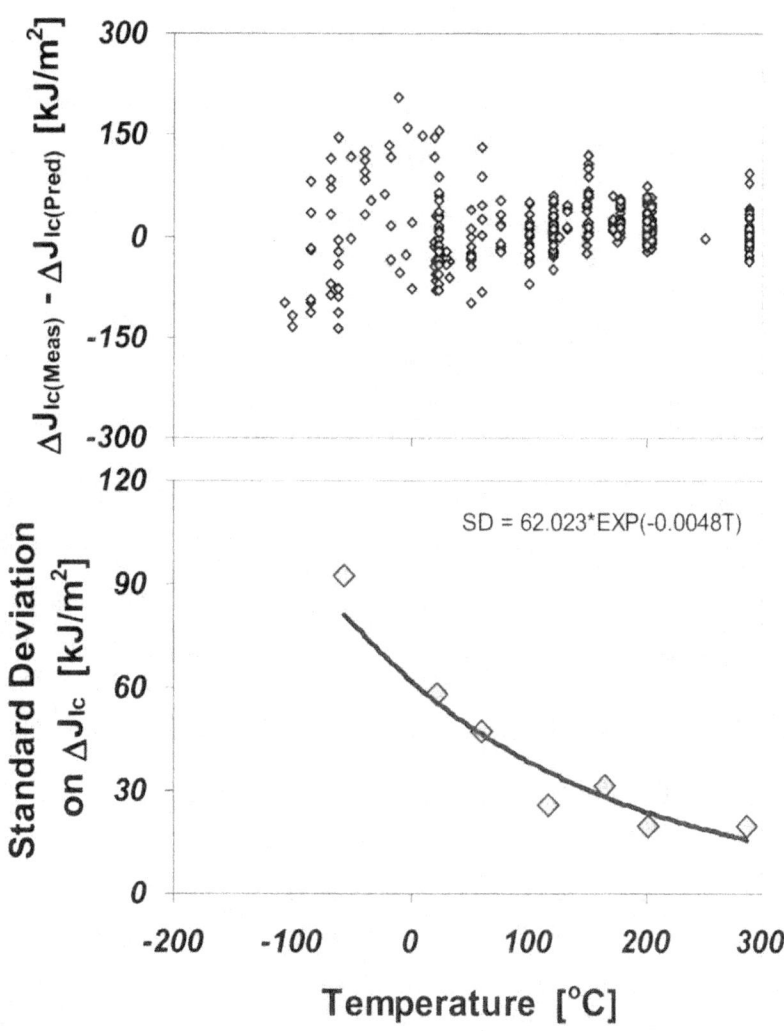

Figure 5-19 (Top) Residuals of measured ΔJ_{Ic} values about the fit of Eq. 5-7, and (bottom) standard deviations fit to these data over discrete temperature ranges. The standard

deviation values are plotted at the midpoint of the temperature range over which the standard deviation was calculated.

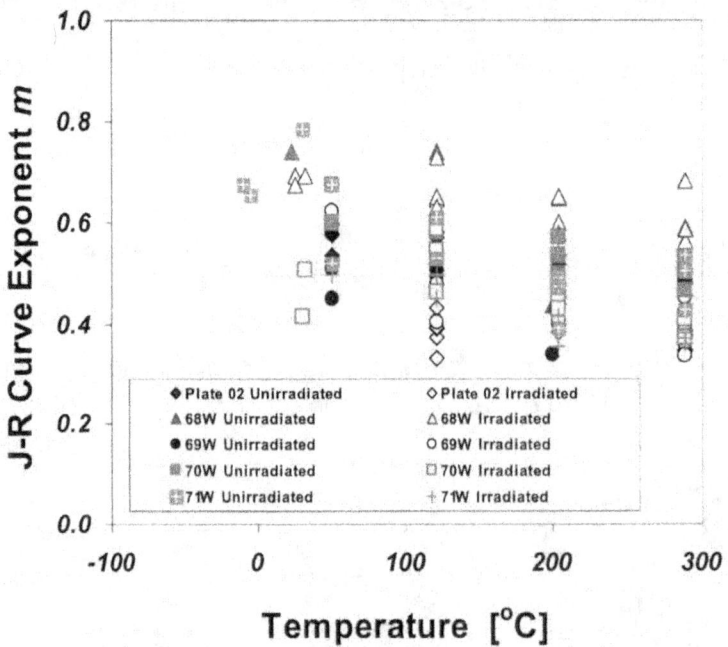

Figure 5-20 Temperature dependency of J-R curve exponent *m* values for a number of nuclear RPV steels (McGowan 88)

In the absence of any physical insight to suggest the technical superiority of one method over the other, FAVOR adopted method 2 for the purpose of computational efficiency.

5.3.1.2   Influence of K_{Ia} on Subsequent K_{Ic} Values

Crack initiation toughness is undefined at values below the crack arrest toughness because the fact that the crack has arrested means that it could no longer propagate. This idea is supported by the experimental observation that the crack arrest toughness transition curve always falls below the crack initiation toughness transition curve. This physical argument establishes the value of crack arrest toughness ($\tilde{K}_{Ia}$) as the minimum allowable K_{Ic} for the temperature at which crack arrest occurred, $\tilde{T}$. As the transient continues after crack arrest, the temperature at the arrest location falls below $\tilde{T}$. Consequently, restricting the minimum allowable K_{Ic} to $\tilde{K}_{Ia}$ is no longer appropriate. Therefore, for temperatures below $\tilde{T}$, the minimum allowable K_{Ic} therefore cannot exceed the K_{Ia} value of the same percentile as the value $\tilde{K}_{Ia}$ occupies in the K_{Ia} distribution at temperature $\tilde{T}$.

The argument presented in the preceding paragraph establishes the physically admissible bounds on the K_{Ic} distribution presuming that a value of K_{Ia} is known for the material. Were FAVOR modeling crack arrest probabilistically, these bounds would establish the limits of the K_{Ic} distribution. However, FAVOR simulates the aleatory uncertainty in crack arrest data toughness using a Monte Carlo approach in which a large number of deterministic crack arrest analyses are performed to estimate what fraction of the flaws that initiate can be expected to extend through the wall and fail the vessel. In this context, the only consistent choice for K_{Ic} when checking for crack reinitiation is the K_{Ic} value having the same percentile as the K_{Ic} value that initiated the crack originally. This value of K_{Ic} falls within the physically admissible bounds on the K_{Ic} distribution because of the restrictions placed on the K_{Ia} distribution in Section 5.3.1.1.

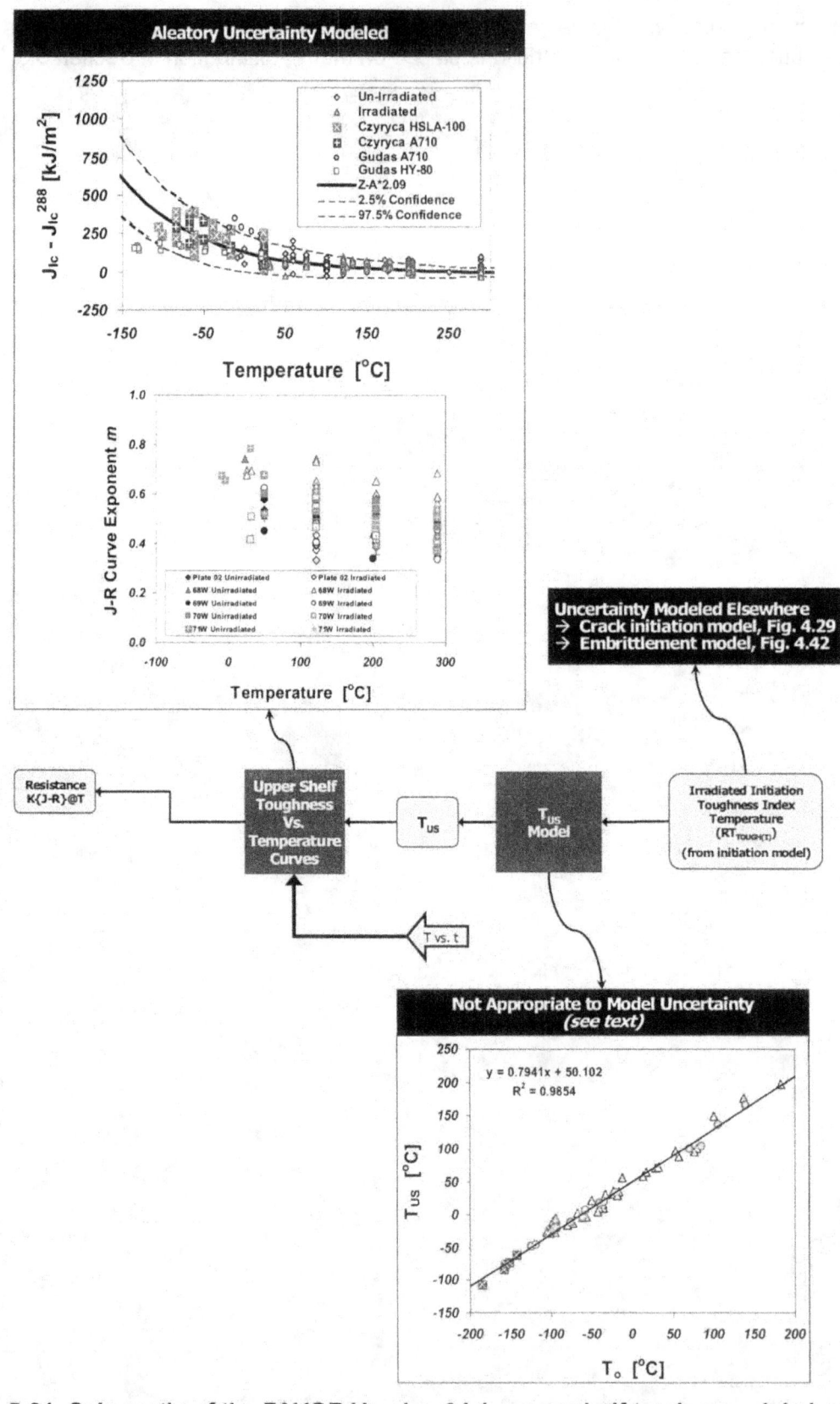

Figure 5-21 Schematic of the FAVOR Version 04.1 upper-shelf tearing model showing the recommended models, input values, and uncertainty treatment

5.3.2 Through-Wall Composition Gradients for Welds

In the early years of nuclear RPV construction in the United States, it was common manufacturing practice to copper coat the welding wires used in RPV fabrication to inhibit corrosion. By the early 1970s, the damaging effects of copper on a steel's resistance to irradiation damage was recognized, and so the practice of copper coating was abandoned. However, all of the early production vessels, those that now lie closest to the PTS screening limits, were fabricated with copper-coated weld wires[‡‡‡]. The copper coating process was not well controlled, which led to varying amounts of copper being deposited on different spools of welding wire. As a consequence of the limited size of these spools and the large volume of weld metal needed to make a PWR weld, it was generally not possible to complete the welding of either the axial seams or the girth welds using a single spool of wire. Evidence of this can be seen in through-thickness compositional surveys (see CEOG). Thus, in order to appropriately model the fracture resistance of the welds in these early vessels, it is important that FAVOR account for the effects of variations in copper content through the thickness of an RPV weld.

Using the following information, one can estimate the number of layers in an RPV weld that can each be expected to have a consistent copper content because the weld metal in the layer was deposited all from the same spool of weld wire:

(1)   the vessel wall thickness

(2)   the vessel diameter

(3)   the dimensions of the weld prep

(4)   the amount of wire in a single spool

(5)   details of the welding process, including whether tandem or single-wire feed was used and information regarding the welding sequence (i.e., how may welds were made simultaneously)

Table 5-6 summarizes this information for the welds in the four plants being modeled, and uses it to estimate (in the last column) the number of distinct layers in these RPV welds. The number of layers was determined by dividing T_{WD} by t_{WALL}, rounding to the nearest integer, and adding 1. Rounding off and adding 1 accounts for the use of less-than-full spools of welding wire at the beginning of fabrication.

The information in Table 5-6 demonstrates that RPV welds can be composed of between two and eight weld layers, each having (potentially) different copper contents. In view of the fact that the calculations in Table 5-6 are merely estimates, and in the interest of computational ease, FAVOR models all welds as having four layers.

To simulate the effect of distinct weld layers on vessel integrity, the following procedure is used in FAVOR:

(1)   Divide the vessel thickness evenly into four weld layers.

[‡‡‡]   The three plants being analyzed as part of the PTS reevaluation effort were all early production vessels, and so were all manufactured using copper-coated weld wire.

(2)   During the course of a crack arrest analysis (see Section 5.1.2), crack propagation is simulated through the vessel wall. When the crack tip passes into a new weld layer, determine new values of chemistry (copper, nickel, and phosphorus) using the mean and standard deviation values associated with the weld region in which the crack is located (formulas in Appendix D describe the distributions of copper, nickel, and phosphorous).

(3)   Determine a new value of irradiation shift using these new composition values and the procedure detailed in Appendix D.

(4)   Because the steel in the new weld layer has different material properties than that in the preceding weld layer, eliminate all restrictions on the K_{Ic} and/or K_{Ia} distribution established based on initiation and arrest events that occurred in the preceding layer (see Section 5.3.1) because the physical rationale that justified these restrictions applies only to the material in which crack initiation and/or crack arrest occurred.

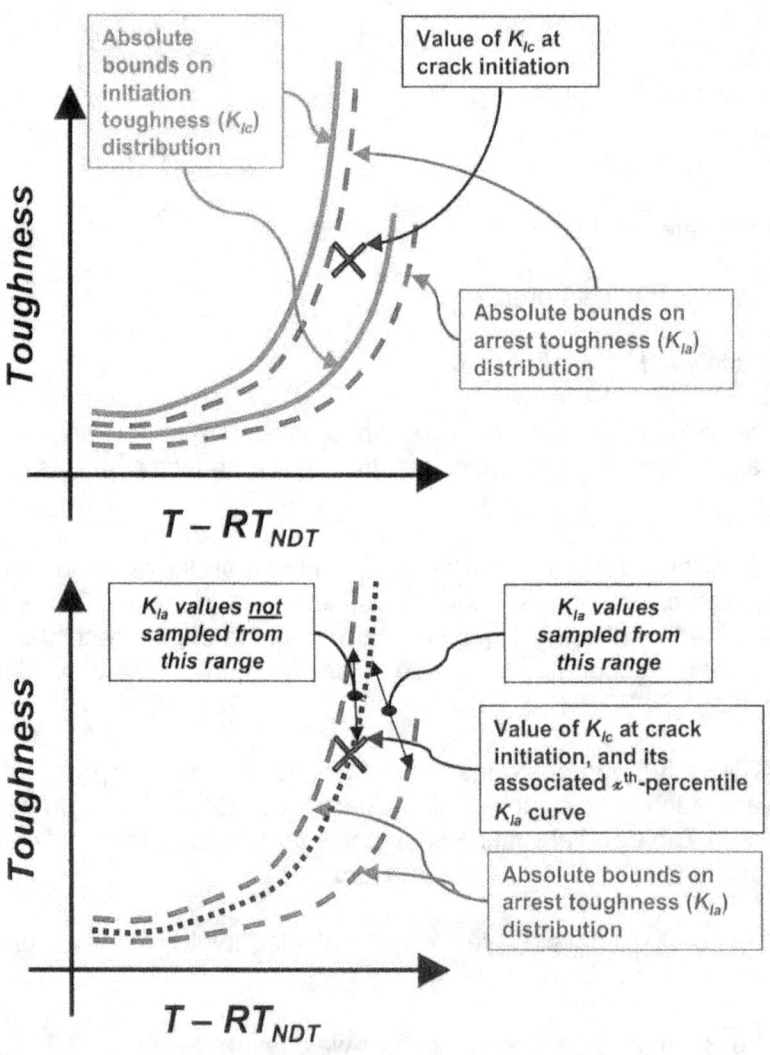

Figure 5-22 Illustration of the proposed procedure to limit K_{Ia} values dependent upon the K_{Ic} value that started the simulation

Table 5-6 Layers of Uniform Copper Content Expected in RPV Welds

Plant	Weld ID	# Welds Made at One Time, N_{WELD}	# of Arcs, N_{ARC}	Vessel Thick., t_{WALL} [in]	Weld Length, L [in.]	Weld Width, W [in.]	Full Spool Weld Layer Thickness, T_{WL} [in.]	Likely Number of Layers in the RPV Weld, N_{WL}
CE Fabrication, Coil Weight (W_{COIL}) = 250 lbs., Coil Volume (V_{COIL}) = 880 cu-in								
Palisades	Intermediate Axial	3	2		96.00	1.4375	4.25	3
	Lower Axial	3	2	8.5	92.72	1.4375	4.40	3
	All Circumferential	1	1		553.71	1.3125	1.21	8
Calvert Cliffs 1	Intermediate Axial	3	2		96.75	1.375	4.41	3
	Lower Axial	3	2	8.625	97.38	1.375	4.38	3
	All Circumferential	1	1		553.90	1.25	1.27	8
Beaver Valley 1	Intermediate Axial	2	2		100.63	1.375	6.36	2
	Lower Axial	2	2	7.875	100.63	1.375	6.36	2
	All Circumferential	1	1		505.60	1.25	1.39	7
B&W Fabrication, Coil Weight (W_{COIL}) = 350 lbs., Coil Volume (V_{COIL}) = 1,234 cu-in								
Oconee 1	Intermediate Axial	2	1		30.00	1.625	12.64	2
	Lower & Upper Axial	2	1	8.44	73.19	1.625	5.18	3
	All Circumferential	1	1		536.40	1.625	1.41	7

Formulas:

$$\rho = 0.284 lbs / in^3$$

$$V_{COIL} = W_{COIL} / \rho$$

$$T_{WL} = \frac{V_{COIL} \cdot N_{ARC}}{N_{WELD} \cdot L \cdot W}$$

$$N_{WL} = ROUND(T_{WL} / t_{WALL}) + 1$$

6 REFERENCES

10 CFR 50.61	*U.S. Code of Federal Regulations*, Section 50.61, "Fracture Toughness Requirements for Protection Against Pressurized Thermal Shock Events," Title 10, "Energy."
Appendix H to 10 CFR Part 50	*U.S. Code of Federal Regulations*, Part 50, Appendix H, "Reactor Vessel Material Surveillance Program Requirements, Title 10, "Energy."
ASME NB2331	American Society of Mechanical Engineers, "1998 ASME Boiler and Pressure Vessel Code, Rules for Construction of Nuclear Power Plants, Division 1, Subsection NB, Class 1 Components," ASME NB-2331.
ASTM E23	American Society for Testing and Materials, "Standard Test Methods for Notched Bar Impact Testing of Metallic Materials," ASTM E23, 1998.
ASTM E185	American Society for Testing and Materials, "Standard Practice for Conducting Surveillance Tests for Light-Water Cooled Nuclear Power Reactor Vessels," ASTM E185-94, 1998.
ASTM E208	American Society for Testing and Materials, "Standard Test Method for Conducting Drop-Weight Test to Determine Nil-Ductility Transition Temperature of Ferritic Steels," ASTM E208, 1998.
ASTM E399	American Society for Testing and Materials, "Test Method for Plane-Strain Fracture Toughness Testing of Metallic Materials," ASTM E399, 1998.
ASTM E900	American Society for Testing and Materials, "Standard Guide for Predicting Radiation-Induced Transition Temperature Shift in Reactor Vessel Materials, E706 (IIF)," ASTM E900-02, 2002.
ASTM E1221	American Society for Testing and Materials, "Standard Test Method for Plane-Strain Crack-Arrest Fracture Toughness, K_{Ia}, of Ferritic Steels," ASTM E1221-96, 1996.
ASTM E1921	American Society for Testing and Materials, "Test Method for Determination of Reference Temperature, T_o, for Ferritic Steels in the Transition Range," ASTM E1921-02, 2002.
Bass 04	Bass, B.R., et al., "Experimental Program for Investigating the Influence of Cladding Defects on Burst Pressure," ORNL/NRC/LTR-04/13, NRC Adams Number ML042660206, 2004.
Bowman 00	Bowman, K.O. and P.T. Williams, "Technical Basis for Statistical Models of Extended K_{Ic} and K_{Ia} Fracture Toughness Databases for RPV Steels," ORNL/NRC/LTR-99/27, Oak Ridge National Laboratory, February, 2000.
Brothers 63	Brothers, A.J. and S. Yukawa, "The Effect of Warm Prestressing on Notch Fracture Strength," *Journal of Basic Engineering*, p. 97, March 1963.

BUGLE	Iskander, S.K., R.D. Cheverton, and D.G. Ball, OCA-I, A Code for Ingersoll 95 (Flu2) Ingersoll, D.T., J.E. White, R.Q. Wright, H.T. Hunter, C.O. Slater, N.M. Greene, R.E. MacFarlane, and R.W. Roussin, "Production and Testing of the VITAMIN-B6 Fine-Group and the BUGLE-93 Broad-Group Neutron/Photon Cross-section Libraries Derived from ENDF/B-VI Nuclear Data," ORNL-6795, NUREG/CR-6214, January 1995.
Carew 01	Private communication.
Charpy	Charpy, M.G., "Note sur l'Essai des Metaux a la Flexion par Choc de Barreaux Entailles," *Soc. Ing. Francais*, p. 848, June 1901.
Cheverton 85a	Cheverton, R.D., et al., "Pressure Vessel Fracture Studies Pertaining to the PWR Thermal-Shock Issue: Experiments TSE-5, TSE-5A, and TSE-6," ORNL-6163, NUREG/CR-4249, Oak Ridge National Laboratory, June 1985.
Cheverton 85b	Cheverton, R.D., et al., "Pressure Vessel Fracture Studies Pertaining to the PWR Thermal-Shock Issue: Experiment TSE-7," ORNL-6177, NUREG/CR-4303, Oak Ridge National Laboratory, August 1985.
Chopra 05	Chopra, O.K., et al., "Crack Growth Rates of Irradiated Austenitic Stainless Steel Weld Heat Affected Zone in BWR Environments," NUREG/CR-xxxx, in publication.
Congleton 85	Congleton, J., T. Shoji, and R.N. Parkins, "The Stress Corrosion Cracking of Reactor Pressure Vessel Steel in High Temperature Water," *Corrosion Science* Vol. 25, No. 8/9, 1985.
Datsco 87	Datsco, J., "Material Properties and Manufacturing Processes," pp. 18–20, John Wiley & Sons, Inc., New York, 1966.
DORT	"DORT, Two-Dimensional Discrete Ordinates Transport Code," RSIC Computer Code Collection, CCC-484, Oak Ridge National Laboratory, 1988.
Eason 91	Eason, E.D., J.E. Wright, and E.E. Nelson, "Multivariable Modeling of Pressure Vessel and Piping J-R Data," U.S. Nuclear Regulatory Commission, NUREG/CR-5729, April 1991.
Eason 98	Eason, E.D., J.E. Wright, and G.R. Odette, "Improved Embrittlement Correlations for Reactor Pressure Vessel Steels," U.S. Nuclear Regulatory Commission, NUREG/CR-6551, 1998.
Eason 03	Kirk, M.T., et al., "Updated Embrittlement Trend Curve for Reactor Pressure Vessel Steels," *Transactions of the 17th International Conference on Structural Mechanics in Reactor Technology (SMiRT 17)*, Prague, Czech Republic, August 17–22, 2003.
EPRI 94	Electric Power Research Institute, "PWR Reactor Pressure Vessel License Renewal Industry Report," Rev. 1, EPRI Report TR-103837, July 1994.
EricksonKirk 04a	EricksonKirk, M., et al., "Technical Basis for Revision of the Pressurized Thermal Shock (PTS) Screening Limit in the PTS Rule (10 CFR 50.61): Summary Report," U.S. Nuclear Regulatory Commission, NUREG-1806.
EricksonKirk 04b	EricksonKirk, M.A., "Materials Reliability Program: Implementation Strategy for Master Curve Reference Temperature, T_o (MRP-101),"

	Electric Power Research Institute, Palo Alto, California, and U.S. Department of Energy, Washington, D.C., 2004. 1009543.
EricksonKirk 04c	EricksonKirk, M., et al., "Sensitivity Studies of the Probabilistic Fracture Mechanics Model Used in FAVOR," U.S. Nuclear Regulatory Commission, NUREG-1808. 2007
English 02	English, C. and W. Server, "Attenuation in US RPV Steels—MRP-56," Electric Power Research Institute, June 2002.
Hurst 85	Hurst, P., et al., "Slow Strain Rate Stress Corrosion Tests on A508-III and A533B Steel in De-Ionized and PWR Water at 563K," *Corrosion Science,* Vol. 25, No. 8/9, 1985.
IAEA 90	*Proceedings of the Third International Atomic Energy Agency Specialists' Meeting on Subcritical Crack Growth,* "Stress Corrosion Cracking of Pressure Vessel Steel in PWR Primary Water Environments," U.S. Nuclear Regulatory Commission, Moscow, May 14–17, 1990.
Kasza 96	Kasza, K.E., et al., "Nuclear Plant Generic Aging Lessons Learned (GALL)," U.S. Nuclear Regulatory Commission, NUREG/CR-6490, October 1996.
Khaleel 00	Khaleel, M.A., et al., "Fatigue Analysis of Components for 60-Year Plant Life," U.S. Nuclear Regulatory Commission, NUREG/CR-6674, June 2000.
Kirk 98	Kirk, M., et al., "Empirical Validation Of The Master Curve For Irradiated And Unirradiated Reactor Pressure Vessel Steels," *Proceedings of the 1998 ASME/JSME Pressure Vessel and Piping Symposium,* San Diego, California, July 26–30, 1998.
Kirk 01a	Kirk, M.T., M.E. Natishan, and M. Wagenhofer, "Microstructural Limits of Applicability of the Master Curve," 32nd Volume, ASTM STP-1406, R. Chona, ed., American Society for Testing and Materials, Philadelphia, PA 2001.
Kirk 01b	Kirk, M., and M.E. Natishan, "Shift in Toughness Transition Temperature Due to Irradiation: ΔT_o vs. ΔT_{41J}, A Comparison and Rationalization of Differences," *Proceedings of the IAEA Specialists Meeting on Master Curve Technology,* Prague, Czech Republic, September 2001.
Kirk 02a	Kirk, M.T., M.E. Natishan, and M. Wagenhofer, "A Physics-Based Model for the Crack Arrest Toughness of Ferritic Steels," *Fatigue and Fracture Mechanics, 33rd Volume, ASTM STP-1417,* W.G. Reuter, and R.S. Piascik, eds., American Society for Testing and Materials, West Conshohocken, Pennsylvania, 2002.
McCabe	McCabe, D.E., et al., "Evaluation of WE-70 Weld Metal from the Midland Unit 1 Reactor Vessel," U.S. Nuclear Regulatory Commission, NUREG/CR-5736, November 2002.
McGowan 88	McGowan, et al., "Characterization of Irradiated Current-Practice Welds and A533 Grade B Class 1 Plate for Nuclear Pressure Vessel Service," U.S. Nuclear Regulatory Commission, NUREG/CR-4880, 1988.
MESH	"MESH—A Code for Determining the DORT Fixed Neutron Source," Memorandum, M.D. Zentner to J.F. Carew, BNL, August 25, 1981.

Meyers 99	Meyers, M.A and K.K. Chawla, "Mechanical Behavior of Materials," Prentice Hall, Upper Saddle River, New Jersey, 1999.
Nanstad 93	Nanstad, R.K., J.A. Keeney, and D.E. McCabe, "Preliminary Review of the Bases for the K_{Ic} Curve in the ASME Code," ORNL/NRC/LTR-93/15, Oak Ridge National Laboratory, Oak Ridge, Tennessee, 1993.
Natishan 98	Natishan, M.E. and M. Kirk, "A Micro-mechanical Evaluation of the Master Curve," *Fatigue and Fracture Mechanics, 30th Volume, ASTM STP-1360,* K. Jerina and P. Paris, eds., American Society for Testing and Materials, 1998.
Natishan 99a	Natishan, M.E., M. Wagenhoefer, and M.T. Kirk, "Dislocation Mechanics Basis and Stress State Dependency of the Master Curve," *Fracture Mechanics, 31st Symposium, ASTM STP 1389,* K. Jerina and J. Gahallger, eds., American Society for Testing and Materials, 1999.
Natishan 99b	Natishan, M.E. and M. Kirk, "A Physical Basis for the Master Curve," *Proceedings of the 1999 American Society of Mechanical Engineers Pressure Vessel and Piping Conference*, American Society of Mechanical Engineers, July 1999.
Natishan 01	Natishan, M.E., "Materials Reliability Program (MRP) Establishing a Physically Based, Predictive Model for Fracture Toughness Transition Behavior of Ferritic Steels (MRP-53)," Electric Power Research Institute, Palo Alto, California, 2001. 1003077.
NRC MEMO 82	Memorandum from M. Vagans to S. Hanauer (DST/NRR), August 30, 1982.
NRC MTEB 5.2	METB 5-2, Branch Technical Position, "Fracture Toughness Requirements," Rev. 1, July 1981.
NRC LTR 02	Memorandum from Thadani to Collins on "Transmittal of Technical Work to Support Possible Rulemaking on a Risk-Informed Alternative to 10 CFR 50.46/GDC 35," July 31, 2002. (ADAMS Accession No. ML022120660)
ORNL 85a	Oak Ridge National Laboratory, "Pressurized Thermal Shock Evaluation of the Calvert Cliffs Unit 1 Nuclear Power Plant," NUREG/CR-4022, ORNL/TM-9408, for the U.S. Nuclear Regulatory Commission, September 1985.
ORNL 85b	Oak Ridge National Laboratory, "Pressurized Thermal Shock Evaluation of the H.B. Robinson Unit 2 Nuclear Power Plant," NUREG/CR-4183, ORNL/TM-9567, for the U.S. Nuclear Regulatory Commission, September 1985.
ORNL 86	Oak Ridge National Laboratory, "Preliminary Development of an Integrated Approach to the Evaluation of Pressurized Thermal Shock as Applied to the Oconee Unit 1 Nuclear Power Plant," NUREG/CR-3770, ORNL/TM-9176, for the U.S. Nuclear Regulatory Commission, May 1986.
Pellini 63	Pellini, W.S. and P.P. Puzak, "Fracture Analysis Diagram Procedures for the Fracture-Safe Engineering Design of Steel Structures," Welding Research Council Bulletin 88, May 1963.
Pellini 76	Pellini, W.S., "Principals of Structural Engineering Technology," United States Office of Naval Research, Library of Congress Catalogue Number 76-50534, 1976.

RG 1.174	U.S. Nuclear Regulatory Commission, "Format and Content of Plant-Specific Pressurized Thermal Shock Safety Analysis Reports for Pressurized Water Reactors," Regulatory Guide 1.174, 1998.
RG 1.99	USNRC Regulatory Guide 1.99, Revision 2, "Radiation Embrittlement of Reactor Vessel Materials," May 1988.
RELAP 99	RELAP5/MOD3 Code Manual, Volume IV: Models and Correlations, June 1999.
RELAP 01	RELAP5/MOD3.3 Code Manual—Volume III, "Developmental Assessment Problems," December 2001.
Rippstein 89	Rippstein, K. and H. Kaesche, "The Stress Corrosion Cracking of a Reactor Pressure Vessel Steel in High Temperature Water at High Flow Rates" *Corrosion Science,* Vol. 29, No. 5, 1989.
RPVDATA	Griesbach, T.J., and J.F. Williams, "User's Guide to RPVDATA, Reactor Vessel Materials Database," Westinghouse Energy Systems Business Unit, WCAP-14616, 1996.
RVID2	U.S. Nuclear Regulatory Commission Reactor Vessel Integrity Database, Version 2.1.1, July 6, 2000.
RG 1.154	U.S. Nuclear Regulatory Commission, "Format and Content of Plant-Specific Pressurized Thermal Shock Safety Analysis Reports for Pressurized Water Reactors," Regulatory Guide 1.154, 1987.
RG 1.162	U.S. Nuclear Regulatory Commission, "Thermal Annealing of Reactor Pressure Vessel Steels," Regulatory Guide 1.162.
RG 1.190	U.S. Nuclear Regulatory Commission, "Calculational and Dosimetry Methods for Determining Pressure Vessel Neutron Fluence," Regulatory Guide 1.190, March 2001.
Rolfe	Rolfe, S.T. and J.T. Barsom, *Fracture and Fatigue Control in Structures: Applications of Fracture Mechanics,* Second Edition, Prentice-Hall, 1987.
SECY-82-465	U.S. Nuclear Regulatory Commission, "Pressurized Thermal Shock," Commission Paper SECY-82-465, November 23, 1982.
SECY-02-0092	U.S. Nuclear Regulatory Commission, "Status Report: Risk Metrics and Criteria for Pressurized Thermal Shock," Commission Paper SECY-02-0092, May 30, 2002.
Simonen 04	Simonen, F.A., et al., "A Generalized Procedure for Generating Flaw Related Inputs for the FAVOR Code," U.S. Nuclear Regulatory Commission, NUREG/CR-6817, Rev. 1.
Siu 99	"Uncertainty Analysis and Pressurized Thermal Shock, An Opinion," U.S. Nuclear Regulatory Commission, 1999. (ADAMS Accession No. ML992710066)
Sokolov 96	Sokolov, M.A. and R.K. Nanstad, "Comparison of Irradiation Induced Shifts of K_{Jc} and Charpy Impact Toughness for Reactor Pressure Vessel Steels," *ASTM STP-1325,* American Society of Testing and Materials, 1996.
Taylor 34	Taylor, G.I., "The Mechanism of Plastic Deformation of Crystals," *Proc. Roy. Soc.,* Vol. A145, pp. 362–404, 1934.
Wagenhofer 01	Wagenhofer, M. and M.E. Natishan, "A Micromechanical Model for Predicting Fracture Toughness of Steels in the Transition Region,"

	33rd Volume, ASTM STP-1417, Reuter and Piascik, eds., American Society for Testing and Materials, Philadelphia, Pennsylvania, 2002.
Wallin 84a	Wallin, K., T. Saario, and K. Törrönen, "Statistical Model for Carbide Induced Brittle Fracture in Steel," *Metal Science,* Vol. 18, pp. 13–16, January 1984.
Wallin 84b	Wallin, K., et al., "Mechanism-Based Statistical Evaluation of the ASME Reference Fracture Toughness Curve," *5th International Conference on Pressure Vessel Technology*, Vol. II, "Materials and Manufacturing," San Francisco, California, American Society of Mechanical Engineers, 1984.
Wallin 84c	Wallin, K., "The Scatter in K_{Ic} Results," *Engineering Fracture Mechanics,* 19(6), pp. 1085–1093, 1984.
Wallin 84d	Wallin, K., "The Size Effect in K_{Ic} Results," *Engineering Fracture Mechanics,* 22, pp. 149–163, 1985.
Wallin 91	Wallin, K., "Statistical Modelling of Fracture in the Ductile to Brittle Transition Region," Defect Assessment in Components—Fundamentals and Applications, G. Blauel and K.H. Schwalbe, eds., ESIS/EGF9, pp. 415-445, 1991.
Wallin 93a	Wallin, K., "Irradiation Damage Effects on the Fracture Toughness Transition Curve Shape for Reactor Vessel Steels," *Int. J. Pres. Ves. & Piping*, 55, pp. 61–79, 1993.
Wallin 93b	Wallin, K., "Statistical Aspects of Constraint with Emphasis on Testing and Analysis of Laboratory Specimens in the Transition Region," *Constraint Effects in Fracture, ASTM STP-1171,* E.M. Hackett, K.H. Schwalbe, and R.H. Dodds, eds., American Society for Testing and Materials, 1993.
Wallin 97b	Wallin, K., "Loading Rate Effect on the Master Curve T_o," Paper IIW-X-1403-97, 1997.
Wallin 98	Wallin, K., "Master Curve Analysis of Ductile to Brittle Region Fracture Toughness Round Robin Data: The 'EURO' Fracture Toughness Curve," VTT Manufacturing Technology, VTT Publication 367, 1998.
Wallin 98b	Wallin, K., and R. Rintamaa, "Master Curve Based Correlation Between Static Initiation Toughness K_{Ic} and Crack Arrest Toughness K_{Ia}," *Proceedings of the 24th MPA-Seminar*, Stuttgart, October 8–9, 1998.
Wallin 01	Wallin, K., "Low-cost *J-R* curve estimation based on CVN upper shelf energy," *Fatigue and Fracture of Engineering Materials and Structures*, 24, pp. 537–549, 2001.
WRC 175	PVRC Ad Hoc Group on Toughness Requirements, "PVRC Recommendations on Toughness Requirements for Ferritic Materials," Welding Research Council Bulletin No. 175, August 1972.
Williams 04	Williams, P.T. and T.L. Dickson, "Fracture Analysis of Vessels—Oak Ridge, FAVOR v04.1, Computer Code: Theory and Implementation of Algorithms, Methods, and Correlations," U.S. Nuclear Regulatory Commission, NUREG/CR-6854.
Woods 01	Woods, R., et al., "Selection of Pressurized Thermal Shock Transients to Include in PTS Risk Analysis," *IJPVP*, 78, pp. 179–183, 2001.

Zerilli 87

Zerilli, F.J., and R.W. Armstrong, "Dislocation-Mechanics-Based Constitutive Relations for Material Dynamics Calculations," *J. Appl. Phys.*, Vol. 65, No. 5, pp. 1816–1825, March 1987.

APPENDIX A
LEFM VALIDITY

ORNL/NRC/LTR-03/11, Pugh, C.E. and Bass, B.R., "Results from Large-scale Fracture Experiments and Validation of Linear Elastic Fracture Mechanics for Use in PTS Analyses," June 2003.

ORNL/NRC/LTR-03/11

Results from Large-scale Fracture Experiments and Validation of Linear Elastic Fracture Mechanics for Use in PTS Analyses

C. E. Pugh
Oak Ridge Structural Assessments, Inc.
Knoxville, Tennessee

B. R. Bass
Oak Ridge National Laboratory
Oak Ridge, Tennessee

Date Published – June 2003

Prepared for
U. S. Nuclear Regulatory Commission
Office of Nuclear Regulatory Research
Under Interagency Agreement DOE 1886-N653-3Y
NRC JCN No. Y6533

Prepared by
OAK RIDGE NATIONAL LABORATORY
Oak Ridge, Tennessee 37831-8056
Managed and Operated by
UT-Battelle, LLC
For the
U. S. DEPARTMENT OF ENERGY
Under Contract No. DE-AC05-00OR22725

CONTENTS

LIST OF FIGURES

LIST OF TABLES

Results from Large-Scale Fracture Experiments and Validation of Linear Elastic Fracture Mechanics for Use in PTS Analyses

C. E. Pugh and B. R. Bass

ABSTRACT

This is one in a series of reports which document the up-to-date technologies which contribute to the technical bases of the U.S. Nuclear Regulatory Commission's current evaluation of its rule for safety-regulation reactor pressure vessels (RPVs) in commercial light-water power reactors (LWRs) when exposed to pressurized thermal-shock (PTS) conditions. This report documents results from a large number of large-scale vessel experiments that have been conducted to reveal characteristics of fracture behavior of thick-wall pressure vessels under conditions pertinent to PTS scenarios. Those test results are discussed in terms of validating the applicability of linear elastic fracture mechanics (LEFM) to PTS analyses. This discussion is particularly important because the probabilistic fracture mechanics code used in the NRC evaluations is based on the use of LEFM models.

1.   Introduction

The importance of structural integrity of RPVs has been recognized by all stakeholders from the beginning of the nuclear power enterprise in the U.S. Correspondingly, the NRC and the Atomic Energy Commission (AEC) before it have maintained a strong research program to evaluate and improve the technology available for use in RPV integrity assessments. The NRC/AEC research efforts have been continuous and integrated with those of other stakeholders in the nuclear power enterprise. Those efforts began in the mid 1960s, and in the case of materials and structures technology, validation of the applicability of fracture mechanics methods to RPV analyses has been of utmost importance. One measure of the importance of the advancements that have been made is the fact that many ASTM standards (e.g. E-1820 for fracture-toughness measurement and E-1221 for crack-arrest test procedures) and much of the fracture mechanics methodology in Sections III and XI of the ASME Boiler and Pressure Vessel Code are derived from results of this research.

It has also been broadly understood that PTS events can present complex challenges to the structural integrity of an RPV. The NRC/AEC research efforts have methodically and sequentially addressed the various technical factors that influence loadings, material properties, and RPV response under credible scenarios including PTS situations. Therefore, the technology employed today in PTS analysis tools is the result of the progress made over these past decades. This includes the methods employed in the NRC's current efforts to evaluate and potentially revise the PTS rule in 10CFR50.

Within the NRC efforts, the FAVOR computer code is used to perform probabilistic fracture calculations for RPVs exposed to credible PTS scenarios. Although the FAVOR code performs probabilistic analyses, the fracture mechanics computations are made up of many deterministic analyses. The following paragraphs summarize important aspects of the NRC research that has contributed to validating the applicability of linear-elastic fracture mechanics (LEFM) to RPV analyses. The activities discussed were carried out at the Oak Ridge National Laboratory (ORNL) under the NRC-sponsored Heavy-Section Steel Technology (HSST) program. A comprehensive summary of the overall efforts of the HSST program's research through the mid 1980s is given by (Whitman 86).

The integrated experimental/analytical studies of RPV behavior made use of specimens whose sizes have ranged from small to large. The studies of the fracture behavior of large-scale specimens are summarized in this report, and they included three distinct phases of experiments that used thick-wall cylindrical specimens. Those phases sequentially addressed vessels exposed to (1) pressure loads, (2) thermal transient loads, and (3) concurrent pressure and thermal transients, and have been historically referred to as Intermediate Test Vessel (ITV) experiments, Thermal-Shock Experiments (TSEs), and Pressurized Thermal-Shock Experiments (PTSEs). Each of these three phases is discussed below in terms of fracture behavior and their roles in validating the applicability of LEFM to RPV analyses. A total of 22 thick-wall cylinder tests made up these phases, and they were carried out from the early 1970s to the mid 1980s.

2. Intermediate Test Vessel (ITV) Experiments

The ITV experiments were planned in the early 1970s and conducted over the following ten years to demonstrate the fracture behavior of thick-wall vessels in the transition range between frangible and ductile fracture behavior, to verify methods of analysis that could be used to predict the observed behavior, and to examine conservatism in the then current ASME design rules for RPVs. The ITV specimens were pressure vessels that had a 6-in. wall thickness, an outer diameter of 39 in., and a test section length of 54 in. Each ITV was designed and fabricated consistent with the then current edition of the ASME Boiler and Pressure Code, and each had an internal design pressure equal to 9,600 psi. After fabrication, ORNL intentionally flawed the vessels with cracks of prescribed geometries and tested them to failure under internal pressure. ORNL analyses and extensive peer review determined that the 6-in. thickness would produce adequate constraint for fracture analysis validation experiments, and that the constraint and stress states would be representative of those which could occur in a full-scale RPV.

Ten ITV vessels were procured, and three were fabricated with cylindrical nozzles having features typical of light-water reactor vessel penetrations. As an example, Figure A-2.1 shows a schematic of the vessel that was used in test ITV-7. Each ITV test vessel had the same dimensions except for the test area (flaw geometry).

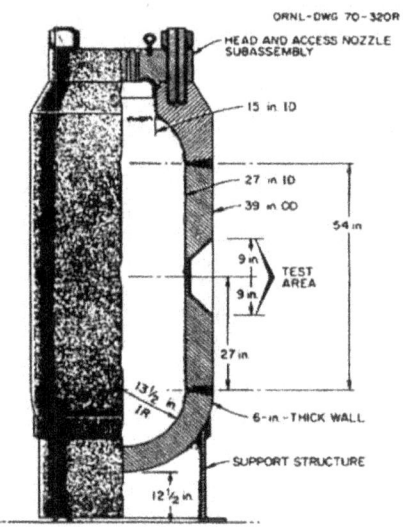

Fig. A-2.1. Schematic of ITV-7 test vessel. Dimensions were the same for each ITV vessel while the specifics of the flaw region and test material varied from test to test.

A total of 12 tests were performed on nine intermediate vessels over a period of 10 years. (One of the nozzle vessels remains untested.) Some of the vessels were used in more than one experiment by replacing the test section with different test materials. Preparation of the tests involved a great deal of engineering development and planning to deliberately fail the vessels within a given range of internal pressures and with high priority given to test safety. Table A-2.1 lists the test dates, materials, test temperatures, flaw sizes, and material toughness values.

Table A-2.1. ITV test sequence and test parameters

Vessel No.	Test Date	Flaw Depth (in.)	Flaw Length (in.)	Test Temp. (°F)	Fracture Toughness (ksi·√in.)	Test Material
1	6/30/72	2.56	8.25	130	311	A508-2
2	9/28/72	2.53	8.30	32	184	A508-2
3	11/8/72	2.11	8.50	130	325	Weld
4	12/20/72	3.00	8.25	75	160	Weld
6	7/7/73	1.87	5.25	190	369	Weld
5	2/13/74	1.20	Nozzle	190	241	A508-2
7	6/19/74	5.30	18.6	196	301	A533-B
9	8/28/74	1.20	Nozzle	75	150-275	A508-2
7A	6/18/76	5.30	18.6	196	301	A533-B
7B	7/14/77	5.30	18.6	196	199-329	Weld HAZ
8	7/26/78	2.50	8.10	-11	90	Weld
8A	8/11/82	2.75	11.0	300	200	LUS Weld

The 12 tests involved cases where the test region was RPV base metal and others where it was weld metal. The test temperatures traversed the Charpy energy curve from the low transition region to well above onset of the upper shelf region. Accordingly, the fracture behavior ranged from brittle fast fracture to total ductile tearing, respectively. This is illustrated in Fig. A-2.2 which shows each test positioned on a Charpy curve. The figure also indicates the nature of fracture observed in these tests.

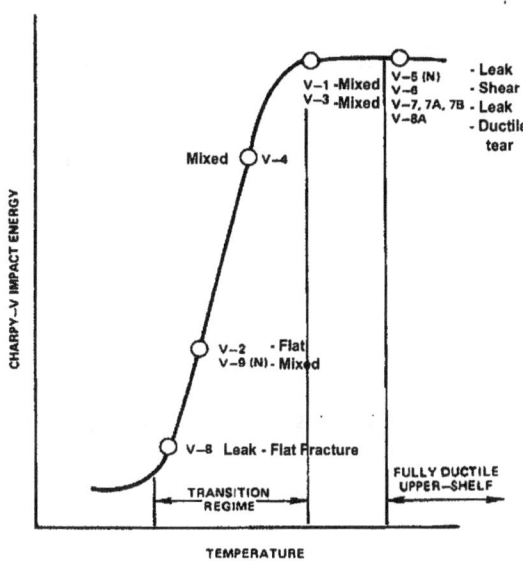

Fig. A-2.2. Test temperatures superimposed on a Charpy curve illustrate relative condition of the twelve ITV experiments. Fracture surfaces are also characterized for each test.

The first test, ITV-1, was performed to show that gross yielding of the vessel wall could be achieved with a large flaw present when the vessel was operating in the upper-shelf regime of toughness. The sequence of tests that followed covered the complete range of fracture failure behavior in relevant materials with varying flaw sizes to include complex stress states and residual stress effects. Table A-2.2 lists the detailed reports that were prepared for each ITV test.

Table A-2.2. List of detailed reports on individual ITV experiments

ITV Test Number	Report Reference	Report Number
ITV-1 and -2	(Derby 74)	ORNL-44895
ITV-3, -4, and -5	(Bryan 75)	ORNL-44895
ITV-5 and -6	(Merkle 77)	ORNL/NUREG-7
ITV-7	(Merkle 76)	ORNL/NUREG-1
ITV-7B	(Bryan 78a)	NUREG/CR-0309 (ORNL/NUREG-58)
ITV-8	(Bryan 78)	NUREG/CR-0675 (ORNL/NUREG-38)
ITV-8A	(Bryan 87b)	NUREG/CR-4760 (ORNL-6187)
ITV-7A	(Bryan 78b)	ORNL/NUREG-9

The range of observed fracture behavior is illustrated by the posttest conditions of test vessels ITV-2 and ITV-6, which are shown in Figs. A-2.3 and A-2.4. Because the temperature for ITV-2 was low in the Charpy transition range (just above NDT) and that for ITV-6 was well above onset of Charpy upper-shelf region, the fracture modes are fully cleavage fast fracture and fully ductile shear, respectively. This range of fracture modes for the thick sections is consistent with fracture behavior observed from tests of smaller laboratory specimens.

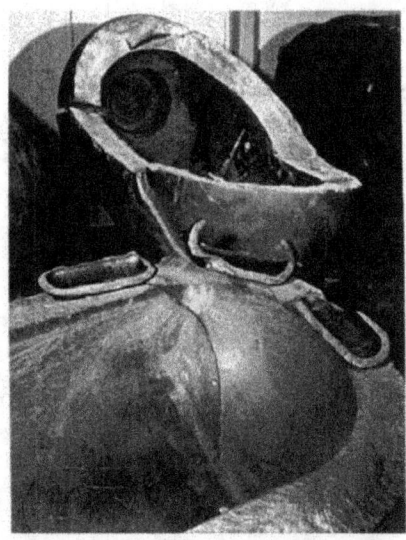

Fig. A-2.3.   ITV-2 vessel was tested under high pressure and at a temperature (32°F) low in the Charpy transition region; fracture was by fast cleavage propagation.

Fig. A-2.4.   ITV-6 test vessel was tested under pressure and at a temperature (190°F) well above onset of Charpy upper shelf; fracture was by ductile shear.

As was the case with each ITV vessel, the ITV-2 vessel contained a fabricated flaw, and because it had a test temperature (32°F) low in the Charpy transition range, it fractured by fast cleavage propagation. It was the only vessel in the ITV series where the crack branched to the point that fragments separated from the remainder of the vessel. All others ITV tests resulted in fracture surfaces whose deviations from flatness and bulging increased from very little to more pronounced (e.g., ITV-6) as the test temperature rose from low to high on the Charpy curve. The flaws that propagated by ductile tearing (e.g., ITV-6) arrested when the crack driving force began to decrease with depressurization due to the through-wall crack.

Most of the other ITV vessels showed mixed cleavage and ductile behavior and the post fracture vessels appeared similar to ITV-1, which is shown in Fig. A-2.5. The fractures were generally flat, and there was limited gross bulging of the vessels. From an overall point of view, the fracture mode results for these tests verified that thick sections of RPV steels exhibit a fracture-mode transition with temperature consistent with that observed for small specimens. In that transition, the fracture goes from brittle cleavage fracture to ductile tearing as the temperature increases from the lower shelf region to the upper shelf region of the Charpy energy curve.

Fig. A-2.5.   ITV-1 vessel was tested under pressure and at a temperature (130°F) high in the Charpy transition region; fracture was mixed cleavage and ductile behavior.

(Merkle 75) provides a thorough evaluation of ITV test results including interpretations relative to Section III, ASME-BPV code. As taken from (Merkle 75), Table A-2.3 shows a summary of the test results for six ITV experiments that used vessels without nozzles. That table also includes comparisons of the experimentally observed failure pressures to those predicted by LEFM analyses and those allowed by ASME design code. Figure A-2.6 graphically displays these results.

Table A-2.3. Results from six ITV experiments using specimens without nozzles

ITV Test No.	P_f Max. Test Pressure (ksi)	P_d ASME Design Pressure (ksi)	P_{LEFM} Predicted Max. Pressure (ksi)	Load Factor P_f/P_{LEFM}	Load Factor P_f/P_d
2	27.9	9.6	27.4	1.018	2.87
4	26.5	9.6	26.2	1.012	2.73
1	28.8	9.6	27.5	1.047	2.96
3	31.0	9.6	27.5	1.127	3.19
6	31.9	9.6	27.5	1.160	3.28
7	21.4	9.6	20.8	1.029	2.20

The data in Tables A-2.2 and A-2.3 show that (in addition to temperature) flaw size and stress concentrations have strong influences on the conditions at failure. This is illustrated by comparing the results of tests ITV-6 and ITV-7. These experiments were conducted at similar temperatures, but ITV-7 had a much deeper flaw (a = 5.3 in.) than did ITV-6 (a = 1.87 in.). Correspondingly, ITV-7 had a lower failure pressure. However, one of the principal objectives of ITV-7 experiment was to demonstrate whether or not a vessel with a very deep flaw could sustain pressures in excess of the design pressure prior to crack initiation. Table A-2.3 and Fig. A-2.6 show that even with the very deep flaw ITV-7 sustained more than two times the design pressure before fracture occurred. Overall, Table A-2.3 and Fig. A-2.6 show that each of the other ITV vessels without nozzles did not exhibit fracture until the applied pressure exceeded the design pressure by a factor of about three [i.e., $(P_f/P_d) \approx 3$].

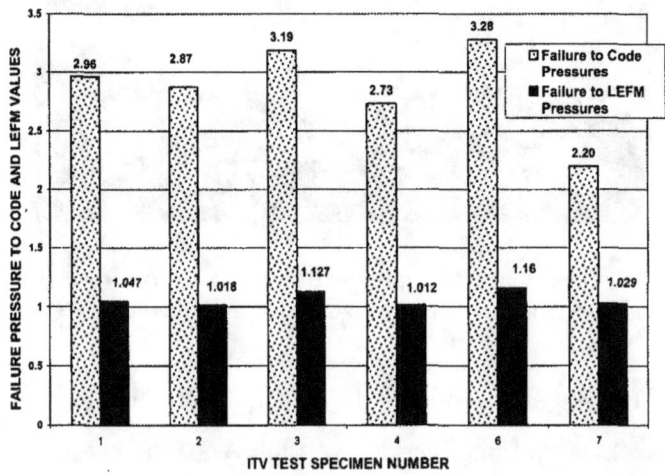

Fig. A-2.6. Load factors for six large-scale ITV experiments [$(P_f/P_d$ = fracture pressure to design pressure ratio) and (Pd/P_{LEFM} = fracture pressure to LEFM predicted pressure ratio)].

Each ITV experiment was analyzed using methods based on LEFM. Table A-2.3 tabulates the failure pressures predicted by LEFM methods (P_{LEFM}) for the ITV vessels without nozzles, and these values are

A-14

seen to be very close to P_f, the test pressures when fracture occurred. This was the case even for ITV-7, which had the very deep flaw. The importance of this excellent agreement is highlighted by the fact that these tests involved such a variety of RPV materials, flaw sizes, and test temperatures. The LEFM analyses made use of fracture and other properties determined from small laboratory specimens made of the respective test materials. Thus, in addition to demonstrating the applicability of LEFM to thick wall pressure vessels, the ITV tests also demonstrated transferability of information from small specimens to large vessels. (Although they are not included in Table A-2.3, the actual failure pressures for the vessels with nozzles were also well predicted by LEFM methods. It is known that flaws at nozzle corners experience only limited constraint, and, they consequently exhibit an equivalent high toughness behavior.)

While these experiments demonstrated the very good agreement between LEFM methods predictions and failure pressures for thick-wall vessels, they also provided opportunities to compare loading conditions allowed by the ASME code with actual failure conditions of those pressure vessels. (Merkle 75) conducted a methodical study that illustrated the conservatism in LEFM methodology even when it was assumed that a flaw might exist with depth equal to the maximum allowed by Section XI, ASME BPV code. In that case, (Merkle 75) showed that the vessel would still experience gross yielding prior to reaching fracture initiation.

In summary, the principal observations and conclusions made from the ITV tests include: (1) the transitional character of fracture-failure was verified for thick-section structures, (2) analyses based on LEFM very closely predicted actual fracture pressures for these thick-wall pressure vessels, (3) flawed vessels fractured at pressure levels significantly greater than the operating pressure allowed under the ASME Code, (4) methods for calculating fracture toughness from small specimens were successfully used in applications of fracture analysis of thick flawed vessels, and (5) the vessels were observed to generally sustain loads three times their design pressure, thus confirming the existence of margins of safety.

3.   Thermal-Shock Experiments

The second phase of HSST large-scale fracture experiments was carried out between 1975 and 1983 and was made up of a set of eight thermal-shock experiments (TSEs) that used thick cylinders made of RPV steel. The purpose of these experiments was to investigate the behavior of surface cracks under thermal-shock conditions similar to those that in principle would be encountered during a large-break loss of coolant accident (LBLOCA). It was known that the injection of cold water by the emergency core cooling system into a hot reactor vessel after a LOCA could produce conditions under which a preexisting flaw might suddenly extend as a result of the low temperature and high thermal stresses. The propensity for such an occurrence is strongly dependent on the degree of degradation in fracture toughness the vessel wall may undergo from radiation damage and the temperature of the cooling water.

The HSST program addressed this issue to ensure that material properties and methods of analysis were developed to a degree that vessel integrity could be accurately assessed throughout the projected vessel lifetime. As the TSE experiments were being planned, ORNL recognized that fracture mechanics analysis for thermal-shock conditions involves several features that had not been adequately examined experimentally at that time. These features included biaxial stresses, steep gradients in stress and toughness through the wall, variations in these parameters with time, crack arrest in a rising K_I field, and warm prestressing (WPS).

To aid in the definition of the TSEs, Fig. A-3.1 was constructed to illustrate through-wall conditions that would exist at a point in time during a thermal shock that would correspond to a LBLOCA. If a flaw of depth "a" were to exist on the inner surface of the vessel of wall thickness "w", the high thermally-induced tensile stress will result in a significant stress-intensity factor at the tip of the flaw. The combined effect of this high K_I and low fracture toughness in the inner-surface region may result in propagation of the flaw. However, the steep positive gradient in the fracture toughness provides a mechanism for crack arrest. Figure A-3.1 includes K_I, K_{IC}, and K_{Ia} curves for this situation and shows that both shallow and deep flaws can initiate. A shallow flaw, which is more likely to exist as an initial flaw, can initiate and propagate through a significant distance before arresting. The deep flaws may be the result of an earlier initiation-arrest event, and at the time shown in this figure, will experience an additional initiation and arrest event.

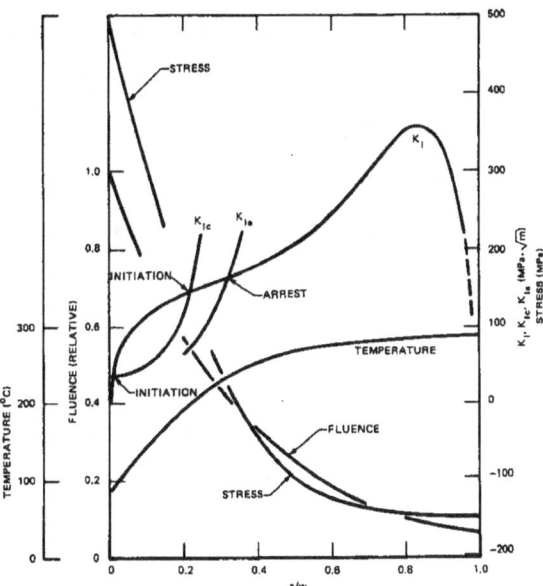

Fig. A-3.1. Typical instantaneous temperature, stress, fluence, stress-intensity factor, and fracture-toughness distributions through wall of PWR vessel during LBLOCA.

ORNL (Cheverton 85a and b) constructed critical-crack-depth curves like those shown in Fig. A-3.2 as an aid in designing and interpreting the TSEs. This type of display shows the predicted behavior of a surface-breaking flaw (of depth a in wall of thickness w) during a specified transient by plotting the crack depths corresponding to initiation and arrest events ($K_I = K_{Ic}$ and $K_I = K_{Ia}$). Multiple crack run-arrest events are shown in this example computation.

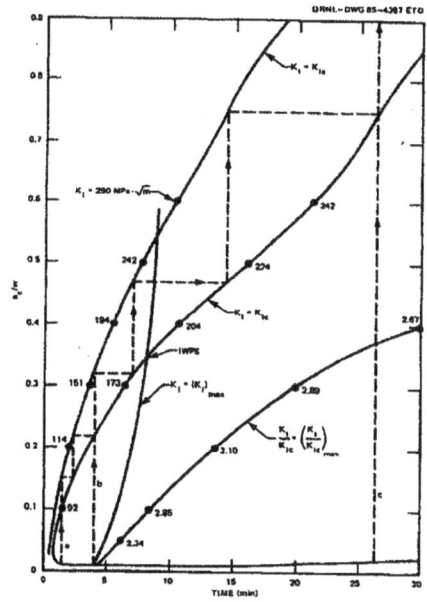

Fig. A-3.2. Critical-crack-depth curves for PWR vessel during LBLOCA.

A-17

These curves illustrate that flaws of different depth could initiate at the same time into the transient and arrest at the same depth. They also illustrate that LEFM analyses predict that a given initial flaw can experience multiple run-arrest events. Similar to observations from Fig. A-3.1, Fig. A-3.2 illustrates that deep flaws that are subject to initiation may exist because it was a preexisting flaw or because it was created by a shallow flaw experiencing earlier run-arrest events. It is also seen that growth of a deep flaw can result in relatively long jumps.

It was also recognized that warm prestressing (WPS) was capable of preventing reinitiation at depths less than the final arrest depth indicated by Fig. A-3.2. The WPS concept means that a flaw will not initiate when K_I is decreasing with time even though K_I may reach or exceed K_{Ic}. Under thermal-shock loading, a deep flaw can conceivably experience it maximum K_I value at a time before the crack tip temperature has decreased enough to make $K_I \geq K_{Ic}$. During the thermal shock, K_I for a given crack depth increases and then begins to decrease with the temperature gradient. However, K_{Ic} continues to decrease as long as the temperature continues to decrease. The curve in Fig. A-3.2 labeled $K_I = (K_I)_{max}$ shows the times at which K_I reaches a maximum for various crack depths. Thus, for times less than indicated by this curve, K_I is always increasing, and for greater times, K_I is always decreasing. If WPS is effective, then crack initiation would be limited to times to the left of this curve. Another feature of the WPS effect would be that if K_I were to later begin to increase (e.g. due to repressurization), the material would exhibit an apparent elevated fracture toughness for temperatures below that for which K_{Ic} is approximately equal to the previously experienced $(K_I)_{max}$.

All these observations contributed to the design of the TSE experiments. The TSE series included scenarios that involved both long and short crack jumps, as well as cases to examine fracture behavior where multiple crack initiation-arrest events would occur and potentially experience intervention by WPS effects. Eight TSE experiments were carried out in two phases that used different specimens and test conditions as summarized below.

The first four TSEs were conducted from September 1975 to January 1977 and used hollow cylindrical specimens fabricated from the trepanned cores taken from the ITV forgings (A508 class-2 steel). These tests are discussed in detail by (Cheverton 76 and 77). The test system used chilled water or water-alcohol mixtures (-23°C) to produce thermal stresses in the heated (288°C) test specimens containing a long internal surface flaw. These test cylinders had an OD of 530 mm (21 in.) and an ID of 240 mm (9.5 in.), and they were 910mm (36 in.) in length. The flaws were shallow with a depth of either 11mm (0.42 in.) or 19 mm (.75 in.). As discussed by (Cheverton 76 and 77), the fracture results from these experiments were in good agreement with predictions based on the LEFM analyses that made use of properties obtained from small laboratory specimens. However, because of the specimen stiffness, deep crack penetrations could not be achieved. Therefore, from August 1979 to May 1983, the second set of four experiments (TSE-5, 5A, 6, and 7) was performed with larger specimens in which deeper crack advance could occur from greater bending, as would be the case in a PWR vessel.

Detailed reports covering this second set of TSE tests are given by (Cheverton 85a and b). The larger cylinders used for these tests were made of A508, Class 2 steel, and had an OD of 991mm (39 in.), ID of 682 mm (27 in.), and length of 1220 mm (48 in.). They were given a quench and temper heat treatment that led to higher (and more conventional) fracture toughness values than was the case for the first four TSE specimens (quench only heat treatment). Thus, a facility that could provide a more severe thermal shock was required for this second set of TSEs. Figure A-3.3 shows a schematic of the liquid nitrogen facility that was constructed at ORNL for testing these larger specimens.

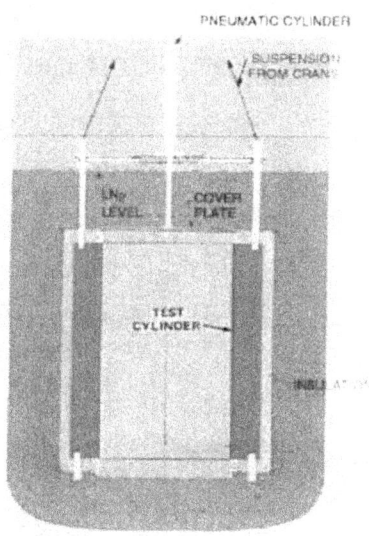

Fig. A-3.3.   Schematic of ORNL's thermal shock test facility used for TSE-5, 5A, 6, and 7.

Figure A-3.4 shows a TSE test cylinder prior to its instrumentation and installation into the TSE test facility. The heated (96°C) cylinders contained inner surface flaws and were submerged into the liquid nitrogen tank (-196°C) to chill their inner surface to produce the desired temperatures and stresses at the crack tips. The test cylinders for TSE-5, 5A, and 6 contained long surface flaws with depths a = 16, 11 , and 7.6 mm, respectively, while the TSE-7 specimen contained a finite length flaw that was 37 mm long and 14 mm deep.

Fig. A-3.4.   TSE test cylinder like those used in TSE-5, -5A, -6, and -7.

TSE-5 experienced a series of three initiation/arrest events with deep penetration (to 80 percent of the wall) of the two dimensional flaw. The occurrence of the three events was entirely consistent with predictions based on LEFM analyses that included the generation of critical crack depth curves [see (Cheverton 85a)]. The analyses predicted a final crack depth of a/w = 0.50 to 0.70 depending upon the effective of WPS effects. Figure A-3.5 shows a posttest cross section of the TSE-5 cylinder that clearly shows the three phases of crack jumps that occurred in this test and led to a final (a/w) of 0.8. Thus, the LEFM analyses predicted well the nature and magnitude of the fracture behavior under this thermal-shock loading.

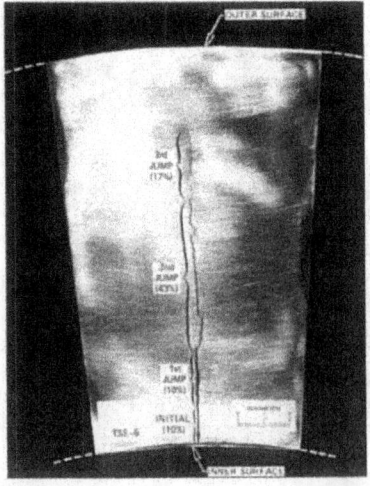

Fig. A-3.5.   Cross section of TSE-5 test cylinder showing the three long crack jumps that occurred during the thermal shock.

As reported by (Cheverton 86), TSE-5A experienced four initiation/arrest events with 50 percent penetration of the wall. Again this was consistent with pretest analyses based on LEFM methods. A fifth event was prevented by WPS effects and one of the arrest events took place with K_I increasing with crack depth. After the WPS intervention, the K_I/K_{Ic} ratio reached a maximum value of 2.3 without crack initiation taking place.

The TSE-6 cylinder had a thinner wall (76 mm vs. 152 mm for the other tests) and introduced the potential for a single long crack jump to a depth greater than 90 percent of the wall thickness. There were actually two crack jumps in the test, the first being relatively short, and the total penetration of 93 percent. This test helped demonstrate the inability of a long flaw to fully penetrate the vessel wall under thermal-shock only loads. According to (Cheverton 85a), this test also demonstrated that there appeared to be negligible dynamic effects associated with arrest following a long crack jump, and the first arrest took place with K_I increasing with crack depth.

TSE-7 was intended to demonstrate the ability of a short and shallow flaw, in the absence of cladding, to extend on the surface to effectively become a long flaw. The initial flaw, which was oriented axially and was essentially semi-elliptical in shape, extended on the surface in a single event and bifurcated many times to produce an extensive cracking pattern [see Cheverton, 85b)]. This event was followed by two crack initiation events that extended the complex flaw to a depth of 55 mm in the central portion of the cylinder and to lesser depths toward the cylinder's ends.

(Cheverton 85a and 86) reported critical values of K_I corresponding to crack initiation and arrest events in these TSE experiments, and his comparisons with laboratory specimen data are shown in Figs. A-3.6 and A-3.7, respectively. (Data from TSE-7 were omitted because of the uncertainty introduced by the complex three-dimensional consideration of the flaws.)

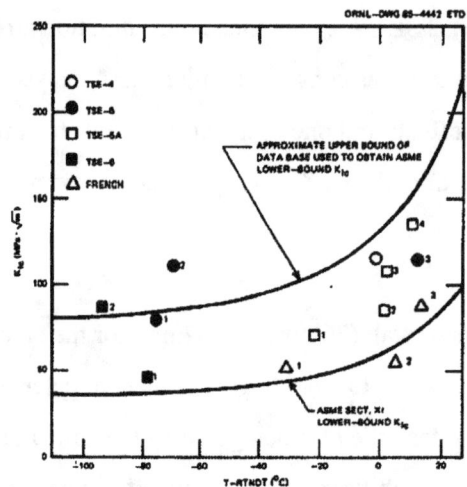

Fig. A-3.6. Comparison of K_{Ic} data from TSE cylinder tests and laboratory specimens.

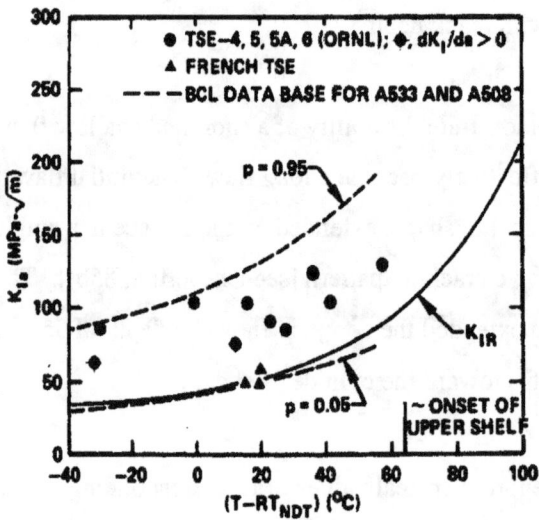

Fig. A-3.7. Comparison of K_{Ia} data from TSE cylinder tests and laboratory specimens.

[The data points (in Figs. A-3.6 and A-3.7) identified as FRENCH data should be ignored in this assessment as discussed by (Cheverton 85b).] The curves shown in these figures are the upper and lower-bound curves from small-specimen data. Overall the K_{Ic} and K_{Ia} values derived from TSE experiments shown in these two figures demonstrate that the behavior of these large-scale fracture situations can be adequately predicted by the use of LEFM methods and fracture properties obtained from tests of small laboratory specimens.

In summary, the objectives of the TSEs were achieved and greatly enhanced the confidence that can be placed in predictive capabilities and the inherent integrity of RPVs. Specific conclusions reached included: (1) multiple initiation-arrest events with deep penetration into the vessel wall were predicted and observed, (2) short, shallow, surface flaws could grow long and deep, (3) WPS limited crack extension through the wall under LOCA conditions, (4) flaws did not penetrate the outer vessel wall even after long crack jumps, (5) small-specimen fracture mechanics data could be interpreted for use in fracture mechanics analysis of thick vessels, and (6) crack arrest could be obtained in a rising stress field.

In addition to these results, the OCA series of computer programs were developed under the thermal-shock project (Iskander 81, Cheverton 84), and these thermal-shock analyses were the forerunner and foundation for the subsequent HSST pressurized thermal-shock tests and evaluations discussed in the next section.

4. Pressurized Thermal Shock Experiments (PTSE)

The third phase of HSST large-scale fracture experiments was composed of two experiments that subjected ITV specimens to concurrent pressure and thermal transients. These PTS experiments were reported in detail by (Bryan 85; Bryan 87a) and represent major milestones in the HSST's long succession of studies relative to fracture prevention for RPVs. These tests have contributed strongly to confirm ting the applicability of theoretical fracture models to the analysis of RPVs. Principal issues of concern in the PTS experiments included:

- crack propagation from brittle to ductile regions;
- crack-propagation under combined time-dependent thermal and pressure stresses
- warm-prestressing effects;
- nature of cleavage crack arrest at temperatures near or above onset of Charpy upper shelf;
- behavior of low upper-shelf energy steels.

In the early 1980s, the HSST program developed a facility for performing PTS-type experiments that exposed intentionally flawed thick-wall pressure vessels to combined thermal and pressure transient loadings. The scale of the tests was chosen to be large enough to attain full-scale constraint of the vessel's flawed region. Test conditions and materials were selected to produce stress fields and gradients around the flaw that are characteristic of RPVs and to provide realistic fracture-toughness conditions. The experiments were designed and analyzed using small-specimen fracture-toughness data and the OCA (Iskander 81, Cheverton 84) computer code. [The OCA (Over-Cooling Accident) code was a precursor to the FAVOR (Dickson 02) code which is used today in the NRC's PTS reevaluation efforts.]

In this report, the focus is on the first PTS experiment performed, i.e., PTSE-1 (Bryan 85). That experiment incorporated a surface crack that was long, sharp and shallow, as assumed in regulatory assessments at the time of the test. The material properties were typical of those for an RPV subjected to moderate neutron embrittlement of the wall. Analytical studies confirmed that the stress levels and gradients around the outside surface flaw in the test vessel provide an acceptable approximation of those occurring in an RPV with a flaw on the inner surface during a postulated PTS event. Recently, the PTSE-1 experiment was re-analyzed using version 02.2 of FAVOR (Dickson 02). Results of that re-analysis, presented in a following section, provide confirmation that cleavage-crack behavior in large-scale thick-walled pressure vessels is reasonably well described by LEFM methodology as embodied in the FAVOR code.

The geometry and dimensions of the ITV are shown in Fig. A-4.1 and in Table A-4.1. A longitudinal plug of specially tempered SA508, Class 2 steel was welded into the ITV V8-A vessel. A sharp outer-surface flaw (1-m-long) was implanted into the plug by cracking a shallow electron-beam weld under the influence of hydrogen charging. Extensive instrumentation was applied to the vessel to provide direct measurements of crack-mouth opening displacement, temperature profiles through the wall, and internal pressure during the transient. The flawed and instrumented vessel was inserted into an outer vessel, which was electrically heated to bring the vessel to the desired uniform initial temperature of about 290°C. (Figure A-4.2 depicts an ITV being lowered into the outer containment vessel at the HSST test facility). The outer vessel also served as a shroud for the PTS transient. A thermal transient was initiated by suddenly injecting chilled water or a methanol-water mixture into the annulus between the test vessel and the outer vessel. The annulus between the two vessels was designed to permit coolant velocities that would produce the appropriate convective heat transfer from the outer surface of the test vessel for a period of about 10 min. Internal pressurization of the test vessel was controlled independently by a system capable of pressures up to about 100 MPa.

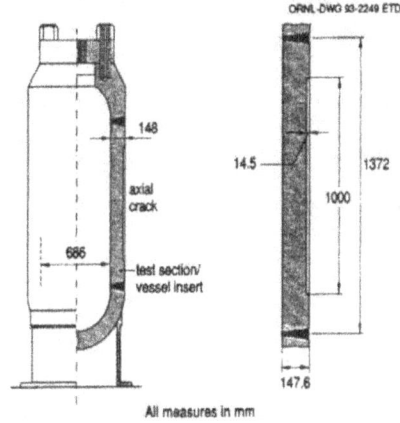

Fig. A-4.1. Geometry of PTSE-1 test vessel.

Table A-4.1. Geometric Parameters of the PTSE-1 vessel

Parameter	Value
Inside radius, mm	343
Wall thickness (w), mm	147.6
Flaw length, mm	1000
Flaw depth (a), mm	12.2
a/w	0.083

Fig. A-4.2. Test vessel being lowered into outer containment vessel at HSST PTS test facility.

Extensive material properties characterization testing and fracture mechanics analyses were carried out during the design phase of the PTSE-1 experiment. Fracture initiation- and arrest-toughness data were generated from tests of 25-mm and 37-mm compact specimens. Those small-specimen data were employed to construct fracture toughness models for planning the test transient. The test plan for PTSE-1 was to achieve initiation and arrest of a fast-running crack, render the arrested crack super-critical (i.e., K_I > K_{Ic}) while in a warm-prestressed (WPS) state, and then re-initiate the crack, driving it toward the completely ductile material deep into the wall. The transient was to be terminated at an appropriate point in time to avoid deliberately rupturing the wall of the test vessel. Figure A-4.3 depicts essential features of the simple WPS cycle envisioned for PTSE-1 experiment. The loading factors of temperature and pressure [Fig. A-4.3(a)] are coordinated to produce a crack driving force K_I that is decreasing with time (dK_I/dt < 0) when the flaw becomes critical [K_I = K_{Ic} in Fig. A-4.3(b)]. Crack propagation of the super-critical crack is inhibited by simple WPS during this period. Warm-prestressing is relieved by increasing the pressure and rendering dK_I/dt >0, thereby introducing the possibility of initiation for the super-critical crack.

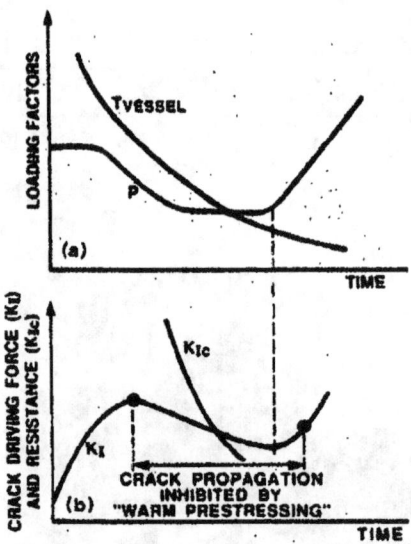

Fig. A-4.3. Features of the simple WPS cycle: (a) loading transient, (b) coordination of transient crack driving force with fracture toughness to induce WPS conditions.

The idealized transient originally designed for the PTSE-1 experiment is depicted in Fig. A-4.4, and the specific objectives were defined for the following intervals.

- Interval A-B: initiation at A and arrest at B of a cleavage fracture;
- Interval B-C: continued loading of the arrested crack;
- Interval C-D-E: crack becomes critical at D, but cleavage initiation is inhibited by WPS;
- Interval E-F: WPS is relieved by increasing the pressure beginning at E;

- Interval F-G: cleavage re-initiation at F and arrest on ductile upper shelf at G;
- Beyond G: termination of transient to prevent through-wall failure.

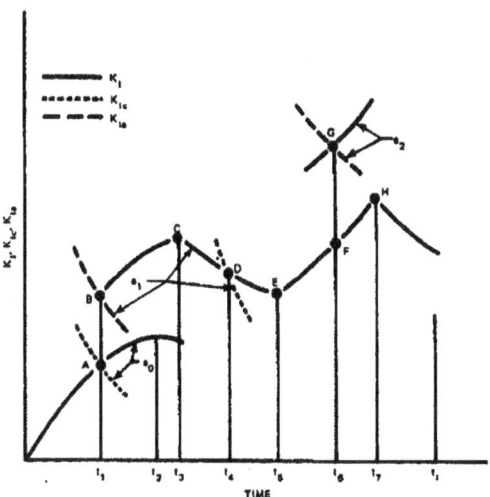

Fig. A-4.4. Idealized crack driving force transient for PTSE-1.

The actual PTSE-1 experiment was conducted in three transients, denoted as PTSE-1A, -1B, and –1C. Initial flaw depths and vessel/coolant temperatures for those transients are summarized in Table A-4.2; time histories of the applied internal pressure are depicted in Fig. A-4.5. The K_I trajectories constructed from experimental data recorded during the three transients are shown in Fig. A-4.6. Because temperature (on the abscissa) decreased monotonically with time, the temporal progression of the K_I trajectories is from right to left in Fig. A-4.6. In the PTSE-1A test, the actual pressure transient varied slightly from what was planned, the crack was slightly deeper than had been estimated, and the actual toughness was higher than had been estimated. As a consequence, the crack did not propagate during the -1A transient. Inspection of the K_I trajectory for PTSE-1A (see Fig. A-4.6) reveals two episodes of simple WPS ($dK_I/dt < 0$), each followed by simple anti-WPS ($dK_I/dt > 0$) when K_I is greater than K_{Ic}. Termination of the transient produced a third episode of simple WPS.

Table A-4.2 Conditions for PTSE-1A, -1B and -1C transients

Transient	PTSE-1A	PTSE-1B	PTSE-1C
Initial vessel temperature (°C)	277.6	290.7	287.4
Coolant temperature (°C)	15 - 34[#]	-22 - 0[#]	-29 - 14[#]
Heat-transfer coefficient (W.m^{-2}K^{-1})	8000-6000[#]	5500-6500[#]	4000-5500[#]
Initial flaw depth, a (mm)	12.2	12.2	24.4
a/w	0.083	0.083	0.165

[#] Initial and final (t ≈ 300s) values

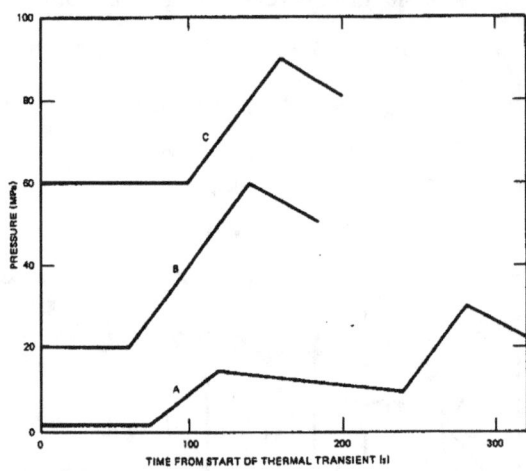

Fig. A-4.5.   Pressure transients for PTSE-1A, -1B, and -1C.

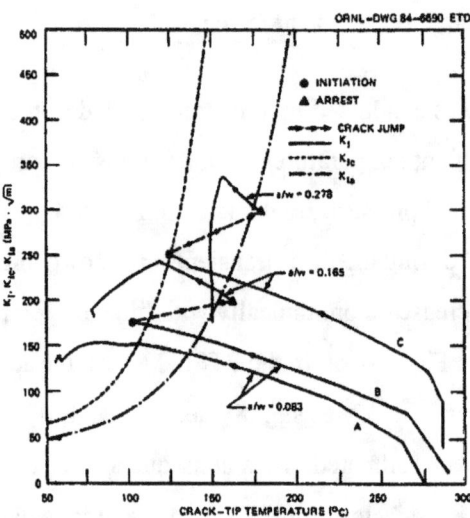

Fig. A-4.6.   Results of OCA analyses of PTSE-1 transients based on measured temperature, pressure, flaw depth, and time of the crack jump.

Plans for the PTSE-1B and -1C transients were based on evidence from test -1A that the vessel had greater toughness than first estimated and that to overcome WPS, a higher K_I value would have to be attained. Thus, a small adjustment was made to the fracture-toughness curve used in pretest analyses, lower coolant temperatures were selected for the thermal transient, and a transient utilizing higher pressure was defined. A two-step pressure transient was not performed during the -1B test because a second pressure increase of useful magnitude was not achievable with the pressurization system. The -1B transient produced a crack jump to a depth of 24.4 mm. The conditions of initiation and arrest are shown in Fig. A-4.6. The arrested crack was subjected to a WPS event in the -1B transient. The third transient,

A-28

PTSE-1C, was subjected to an even higher peak pressure transient, and resulted in a crack jump to a depth of 41 mm under the conditions presented in Fig. A-4.6. Conditions of initiation and arrest depicted in Fig. A-4.6 were determined from analyses performed using the OCA LEFM computer code.

Following completion of the experiment, the flawed region was removed from the test vessel and broken open to reveal the fracture surface. Fractographic examination of the surfaces and measurement of the flaw dimensions indicated that the initial flaw experienced slight tearing prior to the first cleavage event. That initial cleavage run-arrest event (i.e., -1B test) was essentially a pure cleavage fracture for the first half of the extension and primarily cleavage (approx. 90 percent) with finely dispersed ductile tearing in the remainder of propagation. In the second crack jump (i.e., -1C test), crack extension was mixed mode throughout with approximately 85 percent cleavage. No regions of coherent ductile tearing were observed at the ends of the two-crack extension, contrary to the pretest predictions of 2 and 11 mm for transients -1B and -1C, respectively.

At the time of the PTSE-1 experiment, pre- and posttest analyses were carried out using the OCA LEFM computer code. The pretest calculations were based on small-specimen data, while the posttest interpretations utilized data recorded during the experiment. In Fig. A-4.7, K_{Ic} and K_{Ia} values inferred from the experiment are compared with the pretest estimates and with the K_I and K_{Ia} relations from Section XI of the ASME Boiler and Pressure Vessel Code. The pretest estimates of fracture toughness obtained from small specimen data are reasonably close to the measured PTSE-1 values. Also, as expected, the fracture-toughness relations from Section XI are conservative with respect to the measured data.

Recently, the PTSE-1 experiment was re-analyzed by ORNL using version 02.2 of the FAVOR code. The results from FAVOR calculations for the -1B transient are shown in Fig. A-4.8. Analyses based on small-specimen data [see Fig. A-4.8(a)] predict two crack propagation events, the first of which has the flaw initiating immediately after becoming critical (i.e., $K_I / K_{Ia} = 1$) and then arresting at $a = 19$ mm. As depicted in Fig. A-4.8(b), the flaw actually initiated somewhat later in the transient, and arrested at a depth of $a = 24$ mm; the arrested flaw did not re-initiate after becoming critical just before the onset of WPS. These results are interpreted as a re-confirmation that cleavage-crack behavior in large-scale thick-walled pressure vessels is reasonably well described by LEFM methodology as embodied in the FAVOR code.

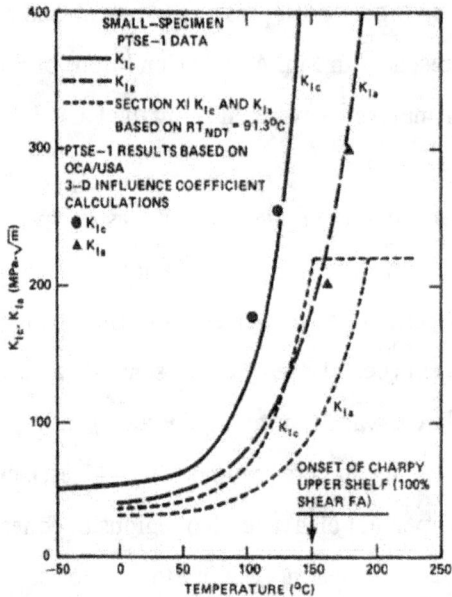

Fig. A-4.7. Comparison of curves representing small-specimen K_{Ic} and K_{Ia} data with ASME Section XI curves and results of PTSE-1 experiment.

Arrest-toughness values measured in the experiment were substantially above the 220 MPa √m cutoff implied in ASME Section XI. Furthermore, Fig. A-4.9 illustrates that the PTSE-1 arrest data are consistent with (1) arrest measurements made in an international set of experiments that include wide-plate and thermal shock tests and (2) the ASME Section XI K_{IR} curve. The highest arrest value recorded in PTSE-1 occurred at a temperature approximately 30° K above the onset of the Charpy upper shelf. These results imply that the methods of LEFM are useful in fracture evaluations of vessels at high Charpy upper-shelf temperatures.

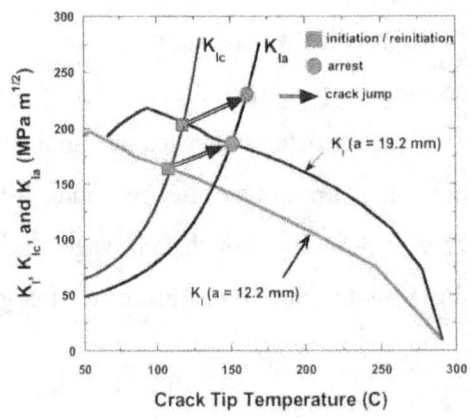

(a)

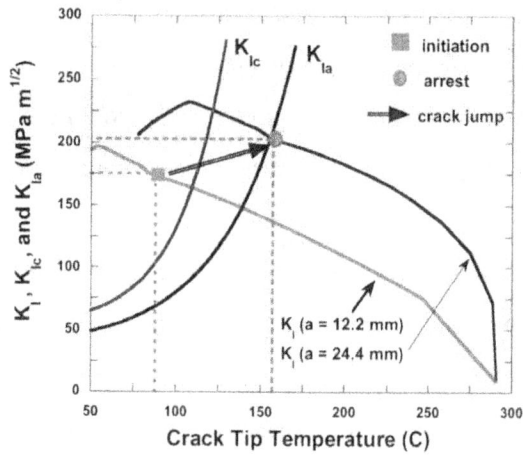

(b)

Fig. A-4.8. Result of FAVOR re-analyses of PTSE-1 experiment: (a) predictions based on small-specimen data; (b) interpretation based on measured temperature, pressure, flaw depth, and time of the crack jump.

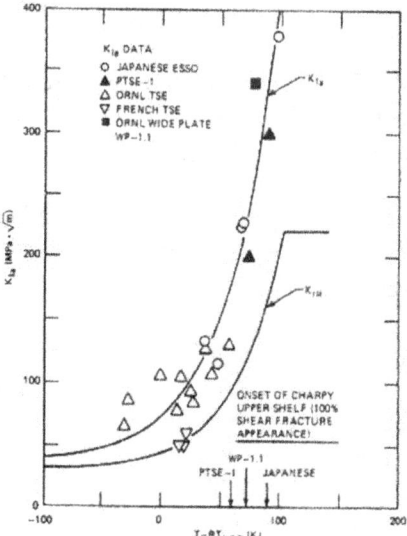

Fig. A-4.9. Comparison of PTSE-1 arrest-toughness results with those for wide-plate and thermal-shock experiments, as well as with ASME Section XI curve.

Both the PTSE-1A and -1B transients demonstrated that simple WPS (i.e., dK/dt < 0) strongly inhibits cleavage crack initiation. In those transients, K_I exceeded K_{Ic} during WPS by 50 to 100 percent without initiation being achieved. In the PTSE-1A transient, simple anti-WPS (i.e., dK/dt > 0) prevailed during two periods of 40-s duration without cleavage initiation, even though K_I exceeded K_{Ic} by 60 to 100 percent. Thus, it can be concluded that simple anti-WPS is not a sufficient condition for overcoming the effects of WPS.

5. Conclusions Based on Large-Scale Experiments

Accomplishments and conclusions supported by results from large-scale HSST experiments (i.e., ITVs, TSEs, and PTSEs) conducted over the past 30-plus years include the following points:

- The fracture behavior observed in large-scale tests for temperatures in the transition range was consistent with the implications of small-specimen data,
- The cleavage-fracture behavior observed in each of the three phases of experiments was well described by LEFM methodology as embodied in OCA/FAVOR computer codes,
- WPS inhibited cleavage-fracture initiation in these experiments where $(dK_I/dt < 0)$, and
- Simple anti-WPS $(dK_I/dt > 0)$ was not a sufficient condition for overcoming the effects of WPS.

6. References

Bryan 75   Bryan, R. H. et al., *Test of 6-inch-Thick Pressure Vessels, Series 2: Intermediate Test Vessels V-3, V-4, and V-6,* Report ORNL-5059, Oak Ridge National Laboratory, November 1975.

Bryan 78a   Bryan, R. H. et al., *Test of 6-Inch-Thick Pressure Vessels. Series 3: Intermediate Test Vessel V-7B,* Report NUREG/CR-0309 (ORNL/NUREG-38), Oak Ridge National Laboratory, October 1978.

Bryan 78b   Bryan, R. H. et al., *Test of 6-Inch-Thick Pressure Vessels. Series 3: Intermediate Test Vessel V-7A Under Sustained Loading,* Report ORNL/NUREG-9, Oak Ridge National Laboratory, February 1978.

Bryan 79   Bryan, R. H. et al., *Test of 6-Inch-Thick Pressure Vessels, Series 3: Intermediate Test Vessel V-8,* Report NUREG/CR-0675 (ORNL/NUREG-58), Oak Ridge National Laboratory, December 1979.

Bryan 85   Bryan, R. H. et al., *Pressurized Thermal Shock Test of 6-in.-Thick Pressure Vessels. PTSE-1: Investigation of Warm-Prestressing and Upper-Shelf Arrest,* Report NUREG/CR-4106 (ORNL-6135), Oak Ridge National Laboratory, April 1985.

Bryan 87a   Bryan, R. H. et al., *Pressurized Thermal Shock Test of 6-in.-Thick Pressure Vessels. PTSE-2: Investigation of Low-Tearing Resistance and Warm-Prestressing,* Report NUREG/CR-4888 (ORNL-6377), Oak Ridge National Laboratory, April 1987.

Bryan 87b   Bryan, R. H. et al., *Test of 6-Inch-Thick Pressure Vessels. Series 3: Intermediate Test Vessel V-8A Tearing Behavior of Low Upper-Shelf Material,* Report NUREG/CR-4760 (ORNL-6187), Oak Ridge National Laboratory, May 1987.

Cheverton 76   Cheverton, R. D., *Pressure Vessel Fracture Studies Pertaining to a PWR LOCA-ECC Thermal Shock: Experiments TSE-1 and TSE-2,* Report ORNL/NUREG/TM-31, Oak Ridge National Laboratory, September 1976.

Cheverton 77   Cheverton, R. D. and Bolt, S. E., *Pressure Vessel Fracture Studies Pertaining to a PWR LOCA-ECC Thermal Shock: Experiments TSE-3 and TSE-4 and Update of TSE-1 and TSE-2 Analysis,* Report ORNL/NUREG-22, Oak Ridge National Laboratory, December 1977.

Cheverton 84   Cheverton, R. D. and Ball, D. G., *A Deterministic and Probabilistic Fracture Mechanics Code for Applications to Pressure Vessels,* Report NUREG/CR-3618 (ORNL-5991), Oak Ridge National Laboratory, May 1984.

Cheverton 85a   Cheverton, R. D. et al., *Pressure Vessel Fracture Studies Pertaining to the PWR Thermal-Shock Issue: Experiments TSE-5, TSE-5A, and TSE-6,* NUREG/CR-4249 (ORNL-6163), Oak Ridge National Laboratory, June 1985.

Cheverton 85b   Cheverton, R. D. et al., *Pressure Vessel Fracture Studies Pertaining to the PWR Thermal-Shock Issue: Experiment TSE-7,* Report NUREG/CR-4303 (ORNL-6177), Oak Ridge National Laboratory, August 1985.

Cheverton 86   Cheverton, R. D., Iskander, S. K., and Ball, D. G., "Review of Pressurized-Water-Reactor-Related Thermal Shock Studies," pp.752-766 in *Fracture Mechanics: Nineteenth Symposium,* ASTM 969, Ed. Thomas A. Cruse, July 1986.

Derby 74   Derby, R. W. et al., *Test of 6-Inch-Thick Pressure Vessels. Series 1: Intermediate Test Vessels V-1 and V-2,* Report ORNL-4895, Oak Ridge National Laboratory, February 1974.

Dickson 02   Dickson, T. L. et al., "Status of the United States Nuclear Regulatory Commission Pressurized Thermal Shock Rule Re-Evaluation Project," *Proceeding of ICONE10, Paper ICONE10-22656, International Conference on Nuclear Engineering*, Arlington, Virginia, April 14-18, 2002.

Iskander 81   Iskander, S. K., Cheverton, R. D. and Ball, D. G., *OCA-I, A Code for Calculating the Behavior of Flaws on the Inner Surface of a Pressure Vessel,* Report ORNL/NUREG-84, Oak Ridge National Laboratory, August 1981.

Merkle 75   Merkle, J. G., *An Evaluation of the HSST Program Intermediate Pressure Vessel Tests in Terms of Light-Water-Reactor Pressure Vessel Safety,* Report ORNL-TM-5090, Oak Ridge National Laboratory, November 1975.

Merkle 76   Merkle, J. G. et al., *Test of 6-Inch-Thick Pressure Vessels. Series 3: Intermediate Test Vessel V-7,* Report ORNL/NUREG-1, Oak Ridge National Laboratory, August 1976.

Merkle 77   Merkle, J. G. et al., *Test of 6-Inch-Thick Pressure Vessels. Series 4: Intermediate Test Vessels V-5 and V-6, with Inside Nozzle Corner Cracks,* Report ORNL/NUREG-7, Oak Ridge National Laboratory, August 1977.

Whitman 86   Whitman, G. D., *Historical Summary of the Heavy-Section Steel Technology Program and Some Related Activities in Light-Water Reactor Pressure Vessel Safety Research,* Report NUREG/CR-4489 (ORNL-6259), Oak Ridge National Laboratory, March 1986.

APPENDIX B
WARM PRESTRESS VALIDITY

Kirk, M., "Inclusion Of Warm Pre-Stress Effects In Probabilistic Fracture Mechanics Calculations Performed To Assess The Risk Of RPV Failure Produced By Pressurized Thermal Shock Events: An Opinion," 2002.

INCLUSION OF WARM PRE-STRESS EFFECTS IN PROBABILISTIC FRACTURE MECHANICS CALCULATIONS PERFORMED TO ASSESS THE RISK OF RPV FAILURE PRODUCED BY PRESSURIZED THERMAL SHOCK EVENTS: AN OPINION

MARK KIRK[†]
Senior Materials Engineer
United States Nuclear Regulatory Commission
Office of Nuclear Regulatory Research
Rockville, MD, 20852, USA
mtk@nrc.gov

Abstract: *The United States Nuclear Regulatory Commission and the commercial nuclear power industry in the United States (operating under the auspices of the Electric Power Research Institute) are in the process of re-evaluating the technical basis of current statutory requirements for the fracture toughness needed by a nuclear reactor pressure vessel to maintain its structural integrity during a pressurized thermal shock (PTS) event. These requirements, currently codified as 10CFR§50.61, state that the RT_{NDT} transition temperature must remain less than 270°F (132°C) for axial welds and plates, and 300°F (149°C) for circumferential welds for the plant to continue in routine licensed operation. These requirements are based on an analysis performed in the early 1980s that contained a number of conservatisms, conservatisms whose re-examination is now appropriate in light of the following factors: technical developments in the areas of probabilistic risk assessment, thermal hydraulics, and fracture mechanics; the current regulatory focus on minimizing <u>overall</u> plant risk; and the economic factors resulting from energy price deregulation in the United States. In this paper we assess the technical basis for including warm pre-stress (WPS) effects in the probabilistic fracture mechanics calculations being performed as part of the PTS rule re-evaluation. The information presented herein demonstrates that inclusion of WPS effects in these calculations is consistent with both theoretical expectations and available experimental evidence and is, therefore, appropriate.*

Keywords: Warm pre-stress, pressurized thermal shock, nuclear reactor, probabilistic fracture mechanics.

[†] The views expressed herein represent those of the author and are not an official position of the USNRC.

1. Background

Warm pre-stress (WPS) effects were first noted in the literature in 1963 [1]. These investigators reported (as have many since them) that the apparent fracture toughness of a ferritic steel can be elevated in the fracture mode transition regime if the specimen is first "pre-stressed" at an elevated temperature. Once a specimen is subjected to a certain $K_{applied}$ and has not failed, the temperature can be reduced and the specimen will remain intact despite the fact that the process of reducing the temperature has also reduced the initiation fracture toughness (K_{Ic} or K_{Jc}) to values smaller than $K_{applied}$. In the past four decades, three mechanisms have been identified as producing (to different extents in different situations) the WPS phenomena [2-4]:

1. Pre-loading at an elevated temperature work hardens the material ahead of the crack tip. The increase of yield strength produced by decreasing the temperature "immobilizes" the dislocations in this plastic zone [5-6]. Consequently, additional applied load is needed for additional plastic flow (and, consequently, fracture) to occur at the lower temperature.
2. Pre-loading at an elevated temperature blunts the crack tip, reducing the geometric stress concentration and making subsequent fracture more difficult.
3. If un-loading occurs between the WPS temperature and the reduced temperature residual compressive stresses are generated ahead of the crack tip. The load applied at the lower temperature must first overcome these residual compressive stresses before the loading can produce additional material damage and, consequently, fracture.

A loss of coolant accident (LOCA) poses a potentially significant challenge to the structural integrity of a nuclear reactor pressure vessel (RPV). During a LOCA, operators must quickly replace the water lost through the breach in the primary system with much colder water held in external tanks to prevent exposure of the reactive materials in the core. The temperature differential between the nominally ambient temperature emergency coolant water and the operating temperature of a pressurized water reactor ($\Delta T = 290°F - 20°C = 270°C$) produces significant thermal stresses in the thick section steel wall of the RPV. These stresses would load cracks in the vessel wall, potentially generating $K_{applied}$ values that exceed the toughness of the RPV material. As illustrated in Figure 1.1, $K_{applied}$ first increases and then decreases as these transients progress, with the time of peak $K_{applied}$ varying depending on both the severity of the transient and the location of the crack in the vessel wall. It is the latter part of the transient when $K_{applied}$ decreases with time that is of interest within the context of WPS. If the $K_{applied}$ value generated by a LOCA were to enter the temperature dependent distribution of initiation fracture toughness values during the falling portion of the transient then the WPS phenomena suggests that crack initiation will not occur even though $K_{applied}$ exceeds the initiation fracture toughness of the material (see Figure 1.2).

To date, probabilistic calculations performed in the United States to assess the challenge to RPV integrity posed by pressurized thermal shock events have not included WPS as part of the PFM model [7-9] for two reasons:

1. TH transients were represented as smooth variations of both pressure and temperature with time. However, data taken from operating nuclear plants demonstrates that actual TH transients are not always so well behaved. This created the possibility that, due to short duration fluctuations of pressure and/or temperature with time, the criteria for WPS might be satisfied by the idealized transient, but not by the real transient it was intended to represent.

2. In the past, the probabilistic risk assessment (PRA) models of human reliability (HR) were not sufficiently sophisticated to capture the potential for plant operators to re-pressurize the primary system as part of their response to a reactor vessel integrity challenge. Since such a re-pressurization would largely nullify the benefit of WPS, it was viewed as non-conservative to account for the benefit produced by WPS within a model that may also ignore the potentially deleterious effects of operator actions.

Our current assessment of the PTS rule features both more realistic representations of the TH transients as well as more sophisticated PRA/HR models that consider explicitly both acts of omission and commission on the part of plant operators. These developments make it appropriate to revisit incorporation of WPS effects into the probabilistic fracture mechanics (PFM) computer code FAVOR (Fracture Analysis of Vessels, Oak Ridge; see [10]), which is used to estimate the effect of a PTS challenge on the RPV.

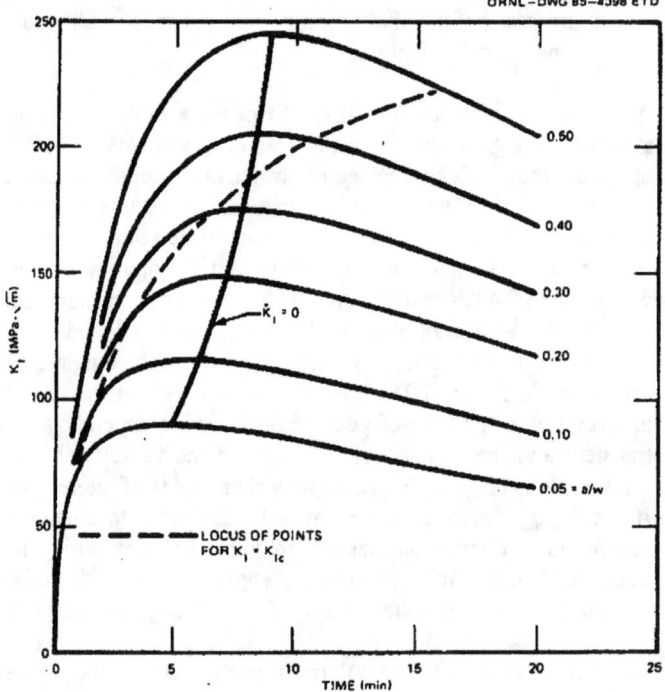

Figure 1.1. Illustration of the influence of crack depth on the variation of $K_{uppplied}$ vs. time resulting from a large break LOCA [11].

The objective of this paper is to determine if sufficient evidence exists to propose a revision to the current FAVOR PFM model, which does not include WPS effects [10], that incorporates the "conservative WPS principal" first proposed by McGowan [12]. This principal states that the criteria for cleavage crack initiation includes not just the commonly accepted requirement that $K_{applied}$ exceed K_{Ic}, but $\underline{also}$ the requirement that $K_{applied}$ must be increasing with time (i.e., $K_{applied}/dt > 0$) when $K_{applied}$ first enters the K_{Ic} distribution. The conservative WPS principal suggests that, even though $K_{applied}$ exceeds K_{Ic}, cleavage fracture cannot occur in the situation depicted by the rightmost diagram in Figure 1.2. Since a number of comprehensive review articles on WPS already exist [2-3] such a review is not repeated here. Rather, in Section 2 we summarize the results of large-scale structural experiments conducted by the NRC in the 1970s and 1980s to assess if the WPS effect is active in RPVs subjected to thermal shock and pressurized thermal shock conditions. On the basis of this summary and other supporting experimental and theoretical evidence we develop a recommended treatment of WPS effects to be incorporated in a future revision of the PFM code FAVOR (see Section 3).

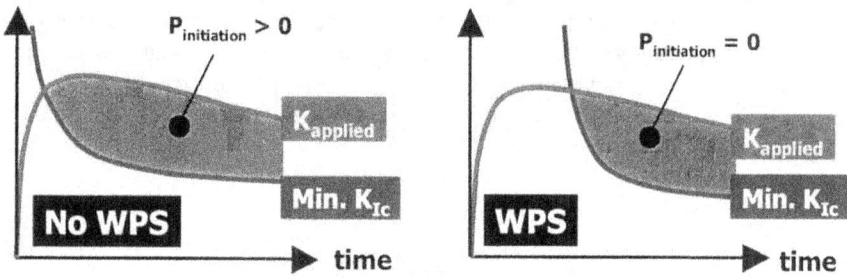

Figure 1.2. Schematic diagram illustrating how the WPS effect could be active during a LOCA depending upon the combination of the transient and the position of the crack within the vessel wall.

2. Evidence of WPS in Large Scale RPV Experiments

In the late 1970s and early 1980s the USNRC Office of Nuclear Regulatory Research sponsored two series of structural-scale RPV experiments at the Oak Ridge National Laboratory under the auspices of the Heavy Section Steel Technology HSST program. The first series of experiments, conducted between 1976 and 1985, focused on the experimental quantification and prediction of the effects of LOCA-type thermal transients on a reactor pressure vessel. The threat of interest during this time was the so-called "large break LOCA." In this transient the postulated break is sufficiently large to rapidly de-pressurize the vessel, so pressure was not a variable modeled in the experiments. On March 20, 1978 Rancho Seco experienced an excessive feedwater transient. Loss of power to control room instrumentation caused operators to maintain reactor coolant system pressure while the vessel was cooled from the operating temperature to 140°C (285°F) in approximately one hour [7]. This event focused attention on the challenges to vessel integrity posed by LOCAs that have less severe thermal stresses (due to smaller break sizes) but during which total de-pressurization cannot be assumed (also due to smaller break sizes) Rancho-Seco was one factor that motivated the conduct of a second series of structural-scale experiments between 1983 and 1989, this time focused on *pressurized* thermal shock events.

Aspects of both the early thermal shock experiments (TSEs) and later pressurized thermal shock experiments (PTSEs) focused on investigating and quantifying the existence of WPS effects. In the following two sections we summarize the experiments that provided evidence of the WPS effect under both TS and PTS conditions. It is also worthwhile to note that none of the experiments conducted in either test series (eight thermal shock experiments and three pressurized thermal shock experiments) provided any evidence that WPS does not occur (i.e., no experiment experienced crack initiation when $K_{applied}$ was falling with increasing time in the transient).

2.1   WPS IN THERMAL SHOCK EXPERIMENTS

In the thermal shock experiments, a thick walled cylinder (nominally 0.9m OD, 1.2m long, having either a 76 mm or 152 mm thick wall) containing either semi-elliptic or uniform depth axial cracks was first heated uniformly, and then chilled rapidly on the inner diameter to initiate cracking. Depending on the particular test conditions a series of initiation / run / arrest / re-initiation (and so on) events ensued. TSE-5 and TSE-5a both exhibited evidence of WPS. Data from these experiments are provided in Figure 2.1 and in Figure 2.2, respectively. In both figures the complete range of K_{Ic} values is superimposed over the part of the transient where WPS may have been responsible for preventing crack initiation, and the portion of this K_{Ic} range that fell below the applied K_I value is cross-hatched.

2.2   WPS IN PRESSURIZED THERMAL SHOCK EXPERIMENTS

In the pressurized thermal shock experiments, a thick walled cylinder (nominally 0.98m OD, 1.3m long, having a 148 mm thick wall) containing a 1m long axial crack of uniform depth was first heated uniformly and then chilled rapidly on the inner diameter to initiate cracking. During this thermal transient pressure also varied, as illustrated in Figure 2.3. In both PTSE-1 and PTSE-2, WPS may have been responsible for the absence of crack initiation during the first of several PTS transients that were applied to each vessel. Data from these experiments are provided in Figure 2.4 and in Figure 2.5, respectively. In both figures the complete range of K_{Ic} values is superimposed over the part of the transient where WPS may have been responsible for preventing crack initiation, and the portion of this K_{Ic} range that fell below the applied K_I value is cross-hatched.

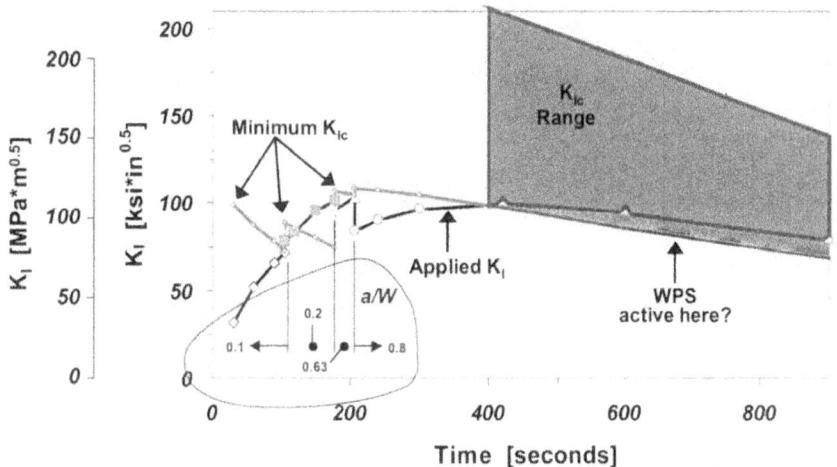

Figure 2.1. Variation of $K_{applied}$ and K_{Ic} with time in TSE-5 showing evidence of a potential WPS effect beginning at ≈400 seconds [11].

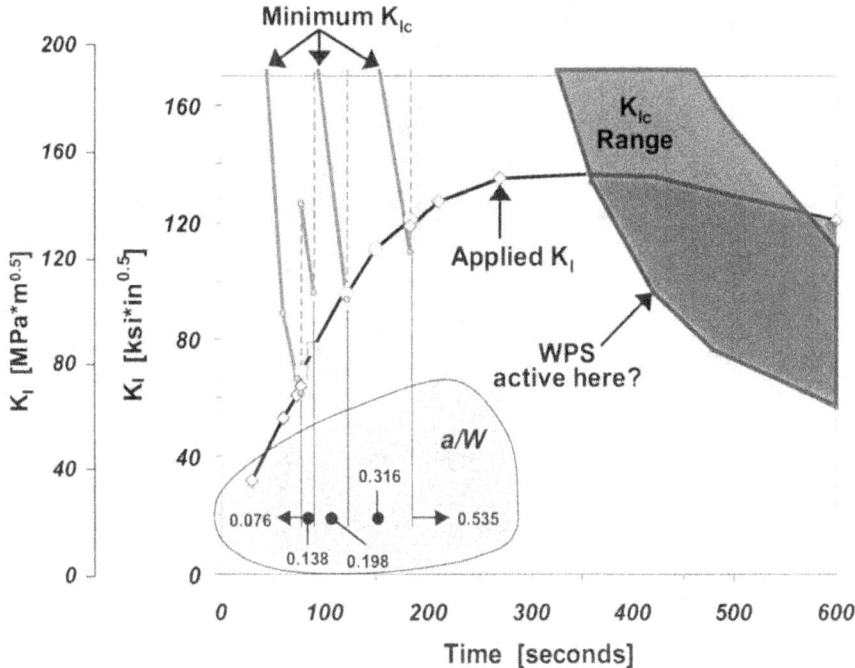

Figure 2.2. Variation of $K_{applied}$ and K_{Ic} with time in TSE-5A showing evidence of a potential WPS effect beginning at ≈360 seconds [11].

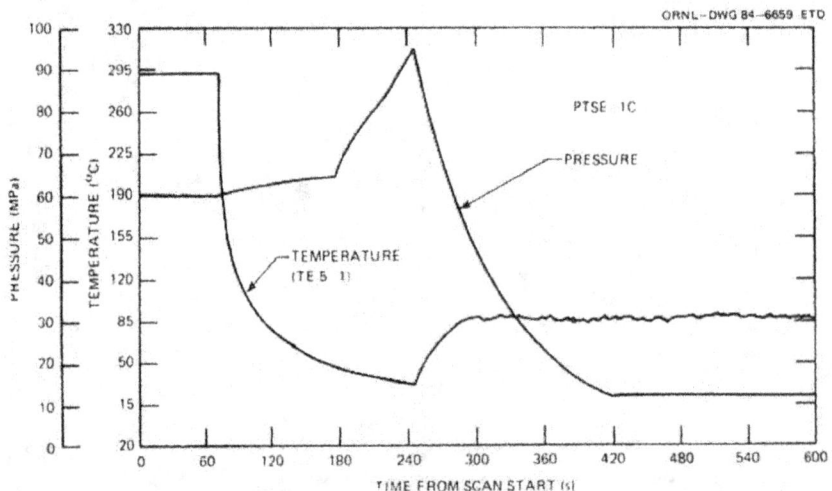

Figure 2.3. Schematic of the pressure / temperature vs. time transients applied during the pressurized thermal shock experiments [13].

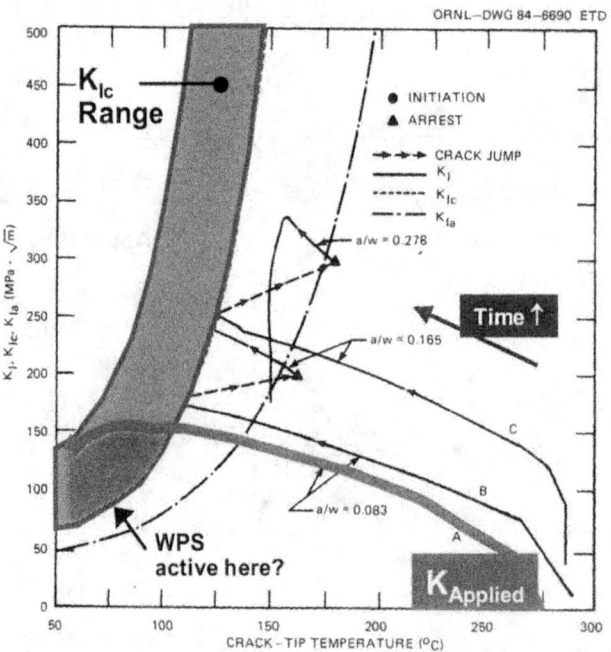

Figure 2.4. Variation of $K_{applied}$ and K_{Ic} with time in PTSE-1 showing evidence of a potential WPS effect in Transient A below a crack tip temperature of $\approx 110\,°C$ [14].

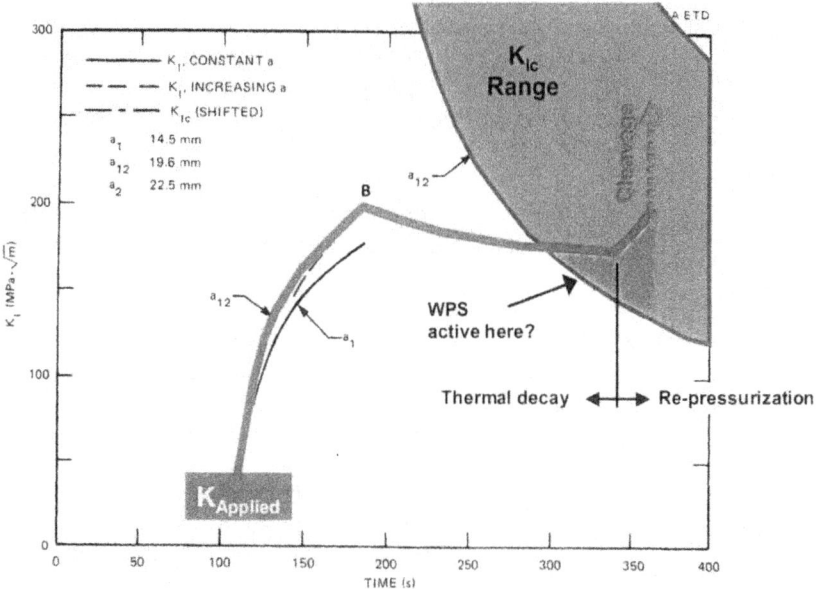

Figure 2.5. Variation of $K_{applied}$ and K_{Ic} with time in PTSE-2 showing evidence of a potential WPS effect beginning at ≈ 300 seconds.

3. Existence of WPS in RPVs Subjected to Thermal Shock and Pressurized Thermal Shock Conditions

3.1 SUMMARY OF EVIDENCE FROM VESSEL EXPERIMENTS

The data summarized in Section 2 demonstrates that in four fracture experiments conducted on prototypic reactor pressure vessels subjected to loadings characteristic of thermal shock and pressurized thermal shock conditions the value of $K_{applied}$ exceeded the minimum value of K_{Ic}, and yet cleavage crack initiation did not occur. In each experiment $K_{applied}$ first exceeded K_{Ic} at a time in the transient when $dK_{applied}/dt$ was either zero or negative, suggesting WPS as one potential explanation for the absence of cleavage crack initiation. However, since the existence of WPS can only be implied based on what does *not* happen (i.e., a cleavage crack does not initiate even though $K_{applied}$ exceeds K_{Ic}), it is prudent to examine other factors that could also explain these observations (see Sections 3.1.1 and 3.1.2).

3.1.1 $K_{applied}$ Is Less Than We Think It Is

As illustrated by the diagram of the crack front for TSE-5 and TSE-5A provided in Figure 3.1 and Figure 3.2 (respectively), the cracks in these experiments took on a decidedly three-dimensional shape because of the reduction in crack driving force near the cylinder's free end. However, the $K_{applied}$ values reported in Figure 2.1 and Figure 2.2 assume a crack of uniform depth equal to the maximum extent of crack penetration into the vessel wall. Relative to this approximation, the correct $K_{applied}$ for the non-uniform depth crack front is lower, suggesting that the crack may not have initiated in these experiments simply because $K_{applied}$ never exceeded K_{Ic}. The impact of this uncertainty on conclusions regarding the existence of WPS in the four structural experiments is as follows:

- <u>TSE-5</u>: Because of the small degree by which $K_{applied}$ exceeded K_{Ic} (see Figure 2.1), it is possible that the Kapplied for the actual (non-uniform depth) crack (shown in Figure 3.1) may not have exceeded K_{Ic}. Thus, some doubt regarding the demonstration of WPS during TSE-5 exists.

- <u>TSE-5A</u>: Uncertainties in $K_{applied}$ are not believed to alter the conclusion that WPS was responsible for the lack of crack initiation in TSE-5A after 360 seconds for two reasons. First, after 180 seconds the crack penetrated to its maximum depth over a length of nearly 0.5m, suggesting that deviations between the $K_{applied}$ values for the crack as it existed in the vessel and the approximate $K_{applied}$ values (estimated by assuming a uniform depth crack of infinite extent) should be small. Furthermore, $K_{applied}$ exceeded the maximum of the K_{Ic} distribution before the end of the transient, suggesting that (were it not for WPS) cleavage crack initiation should have certainly occurred, yet it did not.

- <u>PTSE-1&2</u>: In both of the pressurized thermal shock experiments WPS may have occurred during the first transient. The crack depth during this transient was the pre-test crack depth, making the uniform depth / infinite extent assumptions made in the calculation of $K_{applied}$ appropriate.

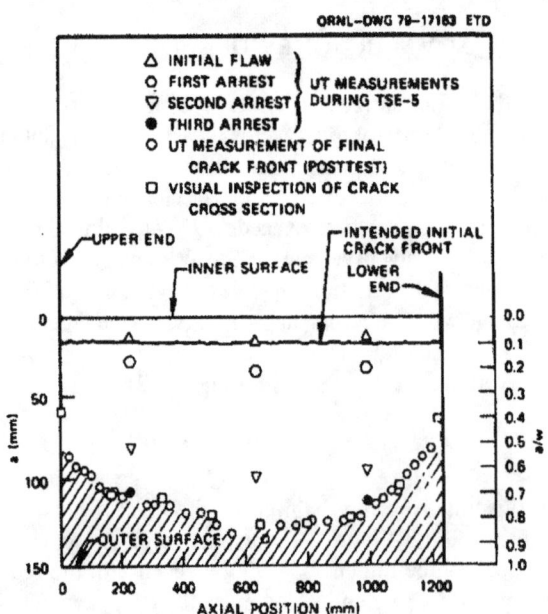

Figure 3.1. Crack profile from TSE-5 [11].

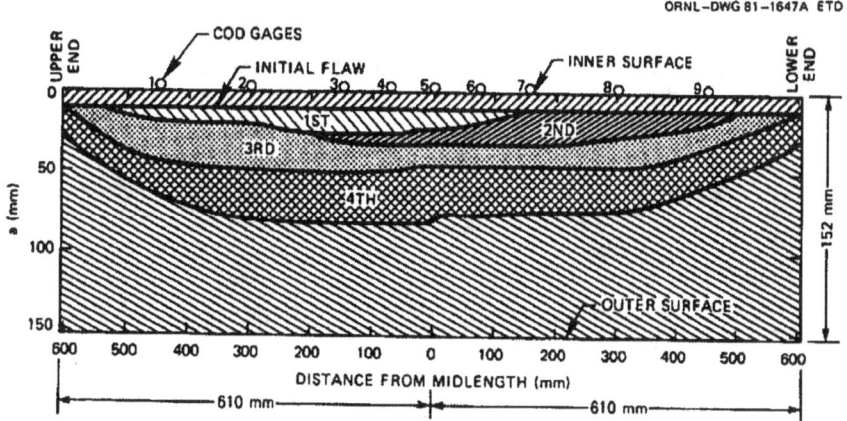

Figure 3.2. Crack profile from TSE-5A[11].

3.1.2 K_{Ic} Exceeds What We Think It Is

Were the K_{Ic} distributions illustrated in Figure 2.1, Figure 2.2, Figure 2.4, and Figure 2.5 for some reason lower than the K_{Ic} for the material at the crack tips in the structural tests this could explain the lack of crack initiation because, in that case, $K_{applied}$ would not have exceeded K_{Ic}. Specifically, the well-documented through-thickness variability in toughness that is expected in rolled plate and extruded forgings could be a confounding factor in this regard [15]. This uncertainty is not believed to influence the conclusions drawn about the existence of WPS in any of the structural experiments discussed in Section 2 for the following reasons:

- The K_{Ic} distributions drawn in these figures is based on fracture experiments conducted using specimens removed from the TSEs and PTSEs themselves, making these material properties the most relevant to understanding the results of the structural test.
- In the TSEs at the time of potential WPS, the crack had advanced well into the portion of the vessel wall thickness where uniform toughness properties are normally observed.
- In the PTSEs the 150 mm thick test vessel was machined from a thicker (203 mm) forging. This forging thickness was reduced to the 150 mm thickness of the PTSEs by machining 38mm from the outer diameter and 13mm from the inner diameter. Thus, even though the crack depth at the time of WPS was shallow (a/W≈0.1) in both experiments, the crack-tip was actually located at deeper into the thickness of the original forging, a region that typically exhibits uniform fracture toughness properties.

3.1.3 Summary

Even taking into account the various factors described in this Section, there is little doubt that WPS was responsible for the non-initiation of a cleavage crack in both TSE-5a and in PSTE-1 owing to the considerable degree to which $K_{applied}$ exceeded K_{Ic} in each experiment. While TSE-5 and PTSE-2 both suggest the possibility of WPS, the conclusion that WPS was _the_ factor responsible for lack of cleavage crack initiation must, with all factors considered (see Sections 3.1.1 and 3.1.2), be made somewhat more equivocally.

While these results are heartening, they do not by themselves provide an adequate technical basis to justify inclusion of WPS in the FAVOR code. Evidence supporting WPS therefore needs to be drawn from other sources (e.g., experimental evidence obtained from specimen tests, and from the theoretical understanding of the WPS phenomena itself: see Section 3.2). Additionally, it is important to recognize that none of these experiments (nor any other experiments conducted to date on either vessels or fracture specimens) have been performed using irradiated materials. Since the aim of this paper is identification of a WPS model that can be applied to irradiated materials, this will be discussed in Section 3.2 as well.

3.2   SUMMARY OF OTHER EVIDENCE

3.2.1   Experimental

Since experiments on fracture toughness specimens can be conducted more economically than prototypic vessel experiments, such results more comprehensively quantify all of the factors relevant to WPS than has been possible using the vessel experiments reported in Section 2. Quoting from a review of warm pre-stressing studies reported by Pickles and Cowan in the *International Journal of Pressure Vessels and Piping* [3],

> *Many experiments have been made on simple fracture toughness specimens to demonstrate that the {WPS} phenomenon exists and, almost without exception, beneficial effects have been found. For cases where no unloading is involved, no reported instance has been found of a specimen failing at low temperature following warm pre-stress without addition of further load above the warm pre-stress load; this is the case despite the fact that the warm pre-stress load could be well above the load to achieve the low temperature {minimum} K_{Ic}.*

Since the no-unloading case represents the upper-bound to $dK_{applied}/dt < 0$ (i.e., $dK_{applied}/dt = 0$), the experimental evidence provides strong testament to the appropriateness of the "conservative warm-prestressing" principal expressed by McGowan that is being considered here for inclusions in a future version of FAVOR [12]. However, since no WPS experiments have been conducted on irradiated materials, the appropriateness of WPS in this situation must be justified on a basis that includes more than just experimental evidence (see Section 3.2.2).

3.2.2 Theoretical

Returning to the three mechanisms of WPS identified in Section 1 we see that the first WPS mechanism involves the effect that pre-loading at an elevated temperature has on work hardening the material ahead of the crack tip. The increase of yield strength produced by decreasing the temperature "immobilizes" the dislocations in this plastic zone [5-6]. Consequently, additional applied load is needed for additional plastic flow (and, consequently, fracture) to occur at the lower temperature. Combining this WPS mechanism with a dislocation-mechanics based understanding of the combined effects of temperature and irradiation on flow properties provides assurance that the "conservative WPS principal" can be *expected* to apply to irradiated steels, even in the absence of direct experimental evidence. Natishan, et al. point out that irradiation influences only the long-range barriers to dislocation motion in ferritic steels, it has no effect on the short-range barriers (provided by the lattice spacing) that control the temperature dependency of the flow behavior [16]. This understanding, combined with an experimentally validated dislocation mechanics based flow model [17] (see Figure 3.3) demonstrates that the increase of yield strength with decreasing temperature needed to ensure the existence of WPS in irradiated materials can be expected on firm theoretical grounds.

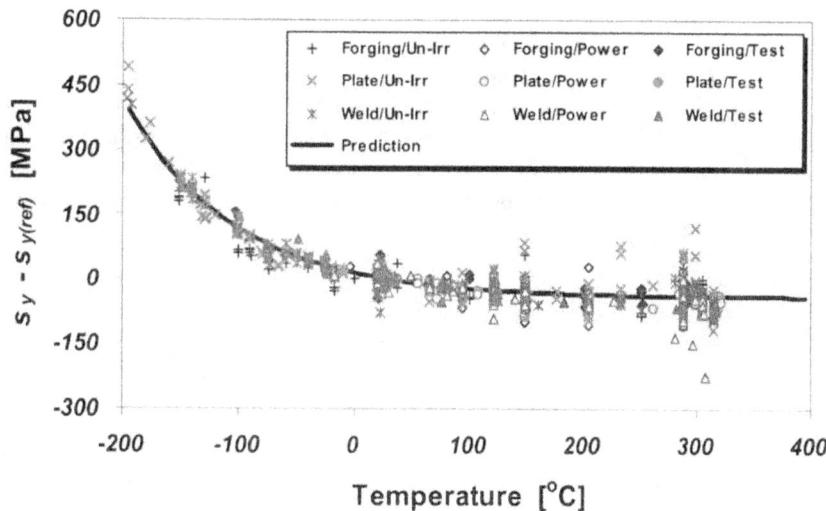

Figure 3.3. Agreement of the thermal component of yield strength in irradiated and un-irradiated RPV steels (irradiations conducted in both test and commercial power reactors) with the dislocation mechanics model (curve labeled "prediction") of Zerilli and Armstrong [17] reported by Kirk, et al. [18].

3.3 RECOMMENDATIONS FOR FAVOR CALCULATIONS

Based on the information provided herein, it is justified to include the "conservative WPS principal" in the probabilistic fracture mechanics code FAVOR. Specifically, the conditional probability of crack initiation (CPI) can be non-zero only if **both** of the following conditions are met:

Condition 1. $K_{applied} \geq K_{Ic(min)}$. The time when this condition is first satisfied is designated t_{WPS}

Condition 2. $dK_{applied}/dt > 0$ when Condition 1 is first satisfied (i.e., at t_{WPS}).

If Conditions 1 and 2 are never both satisfied during the course of a transient then either the crack driving force has never exceeded the minimum value of fracture toughness or, even though it has, WPS has occurred. In either case the CPI is, by definition, zero. However, should the following two conditions also **both** be met at some time after t_{WPS}:

Condition A. $K_{applied}$ at the current temperature/time exceeds the $K_{Ic(min)}$ value at t_{WPS}, and

Condition B. $dK_{applied}/dt > 0$ at this same temperature/time.

then CPI can exceed zero because a significant re-pressurization has occurred. In this case all benefits of WPS are lost, and CPI is calculated accordingly.

These checks for WPS will be made during both calculations made to assess if a crack will initiate from a pre-existing defect, and during calculations made to assess if an arrested crack will re-initiate at some later time in the transient. Because the flaw distributions used in these calculations contain mostly small flaws that are placed close to the inner radius of the RPV [19] we expect that the influence of WPS on preventing first initiations to be minimal. However, a considerably greater effect of WPS is anticipated in preventing re-initiations from cracks that have arrested at depths deeper into the vessel wall.

4. References

[1] Brothers, A.J. and Yukawa, S., "The Effect of Warm Prestressing on Notch Fracture Strength," *Journal of basic Engineering*, March 1963, p. 97.

[2] Nichols, R. W., 1968, "The Use of Overstressing Techniques to Reduce the Risk of Subsequent Brittle Fracture: Parts 1 and 2," *British Welding Journal*.

[3] Pickles, B.W. and Cowan, A., 1983, "A Review of Warm Prestressing Studies," *Int. J. Pres. Ves. & Piping*, 14, 95-131

[4] Chen, J.H., Wang, V.B., Wang, G.Z., and Chen, X., 2001, "Mechanism of Effects of Warm Prestressing on Apparent Toughness of Precracked Specimens of HSLA Steels," *Engineering Fracture Mechanics*, **68**, 1669-1689

[5] Chell, G.C., 1979, "A Theory of Warm Prestressing: Experimental Validation and the Implications for Elastic-Plastic Failure Criteria," CERL Lab Note RD/L/N78/79.

[6] Chell, G.C., 1980, "Some Fracture Mechanics Application of Warm Prestressing to Pressure Vessels," 4th International Conference on Pressure Vessel Technology, Paper C22/80, London.

[7] SECY-82-465, 1982, United States Nuclear Regulatory Commission.

[8] D. L. Selby, et al., 1985, Martin Marietta Energy Systems Inc., Oak Ridge National Lab., Pressurized-Thermal-Shock Evaluation of the Calvert Cliffs Unit 1 Nuclear Power Plant, NUREG/CR-4022 (ORNL/TM-9408).

[9] D.L. Selby, et al., 1985, Martin Marietta Energy Systems Inc., Oak Ridge National Lab., Pressurized-Thermal-Shock Evaluation of the H.B. Robinson Nuclear Power Plant, NUREG/CR-4183 (ORNL/TM-9567).

[10] Williams, P. and Dickson, T., 2001, "Fracture Analysis of Vessels – Oak Ridge, FAVOR v01.0, Computer Code: Theory and Implementation of Algorithms, Methods, and Correlations, NUREG/CR-????, in review.

[11] Cheverton, R.D., et al., 1985, "Pressure Vessel Fracture Studies Pertaining to the PWR Thermal-Shock Issue: Experiments TSE-5, TSE-5A, and TSE-6," United States Nuclear Regulatory Commission, Washington, DC, NUREG/CR-4249.

[12] McGowan, J.J., 1978, "An assessment of the Beneficial Effects of Warm Prestressing on the Fracture Properties of Nuclear Reactor Vessels Under Severe Thermal Shock," Westinghouse Electric Company, WCAP-9178.

[13] Bryan, R.H., et al., 1985, "Pressurized-Thermal-Shock Test of 6-in.-Thick Pressure Vessels. PTSE-1: Investigation of Warm Prestressing and Upper-Shelf Arrest," United States Nuclear Regulatory Commission, Washington, DC, NUREG/CR-4106.

[14] Bryan, R.H., et al., 1987, "Pressurized-Thermal-Shock Test of 6-in.-Thick Pressure Vessels. PTSE-1: Investigation of Low Tearing Resistance and Warm Prestressing," United States Nuclear Regulatory Commission, Washington, DC, NUREG/CR-4888.

[15] Viehrig, H.-W., Boehmert, J., and Dzugan, J., 2002, "Some Issues By Using the Master Curve Concept," *Nuclear Engineering and Design*, 212 115-124.

[16] Natishan, M.E., Wagenhoefer, M., and Kirk, M.T., 1999, "Dislocation Mechanics Basis and Stress State Dependency of the Master Curve," *Fracture Mechanics, 31st Symposium, ASTM STP 1389*, K. Jerina and J. Gahallger, Eds., American Society for Testing and Materials.

[17] Zerilli, F. J. and R. W. Armstrong, 1987, "Dislocation-mechanics-based constitutive relations for material dynamics calculations," *J. Appl. Physics*, Vol. 61, No. 5, 1.

[18] Kirk, M.T., Natishan, M.E., M. Wagenhofer, 2001, "Microstructural Limits of Applicability of the Master Curve," *Fracture Mechanics, 32nd Volume, ASTM*

STP-1406, R. Chona, Ed., American Society for Testing and Materials, Philadelphia, PA.

[19] Simonen, F., Schuster, G, Doctor, S., and Dickson, T., 2002, "Distributions of Fabrication Flaws in Ractore Pressure Vessels for Structural Integrity Evaluations," Proceedings of the ASME Pressure Vessel and Piping Conference, Vancouver, British Columbia.

[20] T. J. Burns, et al., 1986, Martin Marietta Energy Systems Inc., Oak Ridge National Lab., Preliminary Development of an Integrated Approach to the Evaluation of Pressurized- Thermal-Shock as Applied to the Oconee Unit 1 Nuclear Power Plant, NUREG/CR-3770 (ORNL/TM-9176).

APPENDIX C
PLANT-SPECIFIC MATERIAL VARIABLES USED IN FAVOR CALCULATIONS

Product Form	Heat	Beltline	$\sigma_{flow(u)}$ [ksi]	$RT_{NDT(u)}$ [°F]		$\sigma_{(u)}$ Value	Composition[2]			$USE_{(u)}$ [ft-lb]
				$RT_{NDT(u)}$ Method	$RT_{NDT(u)}$ Value		Cu	Ni	P	
Beaver Valley 1 (Designer: Westinghouse, Manufacturer: CE) Coolant Temperature = XXX°C (547°F), Vessel Thickness = 7 7/8 in.										
PLATE	C4381-1	INTERMEDIATE SHELL B6607-1	83.8	MTEB 5-2	43	0	0.14	0.62	0.015	90
	C4381-2	INTERMEDIATE SHELL B6607-2	84.3	MTEB 5-2	73	0	0.14	0.62	0.015	84
	C6293-2	LOWER SHELL B7203-2	78.8	MTEB 5-2	20	0	0.14	0.57	0.015	84
	C6317-1	LOWER SHELL B6903-1	72.7	MTEB 5-2	27	0	0.2	0.54	0.01	80
LINDE 1092 WELD	305414	LOWER SHELL AXIAL WELD 20-714	75.3	Generic	-56	17	0.337	0.609	0.012	98
	305424	INTER SHELL AXIAL WELD 19-714	79.9	Generic	-56	17	0.273	0.629	0.013	112
LINDE 0091 WELD	90136	CIRC WELD 11-714	76.1	Generic	-56	17	0.269	0.07	0.013	144
Oconee 1 (Designer and Manufacturer: B&W) Coolant Temperature = XXX°C (556°F), Vessel Thickness = 8 7/16 in.										
FORGING	AHR54 (ZV2861)	LOWER NOZZLE BELT	(4)	B&W Generic	3	31	0.16	0.65	0.006	109
PLATE	C2197-2	INTERMEDIATE SHELL	(4)	B&W Generic	1	26.9	0.15	0.5	0.008	81
	C2800-1	LOWER SHELL	(4)	B&W Generic	1	26.9	0.11	0.63	0.012	81
	C2800-2	LOWER SHELL	69.9	B&W Generic	1	26.9	0.11	0.63	0.012	119
	C3265-1	UPPER SHELL	75.8	B&W Generic	1	26.9	0.1	0.5	0.015	108
	C3278-1	UPPER SHELL	(4)	B&W Generic	1	26.9	0.12	0.6	0.01	81
LINDE 80 WELD	1P0962	INTERMEDIATE SHELL AXIAL WELDS SA-1073	79.4	B&W Generic	-5	19.7	0.21	0.64	0.025	70
	299L44	INT./UPPER SHL CIRC WELD (OUTSIDE 39%) WF-25	(4)	B&W Generic	-7	20.6	0.34	0.68	(3)	81
	61782	NOZZLE BELT/INT. SHELL CIRC WELD SA-1135	(4)	B&W Generic	-5	19.7	0.23	0.52	0.011	80
	71249	INT./UPPER SHL CIRC WELD (INSIDE 61%) SA-1229	76.4	ASME NB-2331	10	0	0.23	0.59	0.021	67
	72445	UPPER/LOWER SHELL CIRC WELD SA-1585	(4)	B&W Generic	-5	19.7	0.22	0.54	0.016	65
	8T1762	LOWER SHELL AXIAL WELDS SA-1430	75.5	B&W Generic	-5	19.7	0.19	0.57	0.017	70
	8T1762	UPPER SHELL AXIAL WELDS SA-1493	(4)	B&W Generic	-5	19.7	0.19	0.57	0.017	70
	8T1762	LOWER SHELL AXIAL WELDS SA-1426	75.5	B&W Generic	-5	19.7	0.19	0.57	0.017	70

Product Form	Heat	Beltline	$\sigma_{flow(u)}$ [ksi]	$RT_{NDT(u)}$ Method	$RT_{NDT(u)}$ [°F] Value	$\sigma_{(u)}$ Value	Composition[2] Cu	Ni	P	$USE_{(u)}$ [ft-lb]
Palisades (Designer and Manufacturer: CE) Coolant Temperature = XXX °C (532 °F), Vessel Thickness = 8-1/2 in.										
PLATE	A-0313	D-3803-2	(4)	MTEB 5-2	-30	0	0.24	0.52	0.01	87
	B-5294	D-3804-3	(4)	MTEB 5-2	-25	0	0.12	0.55	0.01	73
	C-1279	D-3803-3	(4)	ASME NB-2331	-5	0	0.24	0.5	0.011	102
	C-1279	D-3803-1	74.7	ASME NB-2331	-5	0	0.24	0.51	0.009	102
	C-1308A	D-3804-1	(4)	ASME NB-2331	0	0	0.19	0.48	0.016	72
	C-1308B	D-3804-2	(4)	MTEB 5-2	-30	0	0.19	0.5	0.015	76
LINDE 0124 WELD	27204	CIRC. WELD 9-112	76.9	Generic	-56	17	0.203	1.018	0.013	98
	34B009	LOWER SHELL AXIAL WELD 3-112A/C	76.1	Generic	-56	17	0.192	0.98	(3)	111
LINDE 1092 WELD	W5214	LOWER SHELL AXIAL WELDS 3-112A/C	72.9	Generic	-56	17	0.213	1.01	0.019	118
	W5214	INTERMEDIATE SHELL AXIAL WELDS 2-112 A/C	72.9	Generic	-56	17	0.213	1.01	0.019	118

Notes:

(1) Information taken directly from the July 2000 release of the U.S. Nuclear Regulatory Commission's (NRC) Reactor Vessel Integrity (RVID2) database.

(2) These composition values are as reported in RVID2. In FAVOR calculations, these values should be treated as the central tendency of the copper, nickel, and phosphorus distributions detailed in Appendix D.

(3) No values of phosphorus are recorded in RVID2 for these heats. A generic value of 0.012 should be used, which is the mean of 826 phosphorus values taken from the surveillance database used by Eason, et al. to calibrate the embrittlement trend curve.

(4) No values strength measurements are available in PREP4 for these heats (see PREP). A value of 77 ksi should be used, which is the mean of other flow strength values reported in this appendix.

References:

RVID2   U.S. Nuclear Regulatory Commission Reactor Vessel Integrity Database, Version 2.1.1, July 6, 2000.

PREP   PREP4: Power Reactor Embrittlement Program, Version 1.0," EPRI, Palo Alto, CA: 1996. SW-106276.

APPENDIX D
CHEMICAL COMPOSITION UNCERTAINTY DISTRIBUTIONS

While there is considerable information available concerning the composition of the steels used in U.S. nuclear reactor pressure vessel (RPV) construction (see RVID; RPVDATA), the heats for which sufficient information exists and which can be used to estimate the statistical distribution of the chemical composition are considerably more limited. However, such information is required as input to the probabilistic fracture mechanics (PFM) code FAVOR in order to assess the effects of material variability on the probability of vessel failure. This appendix reviews available data sets in which multiple composition measurements have been made on the same heat of steel. These data are then used to derive generic distributions for copper, nickel, and phosphorus. FAVOR assumes these distributions will apply to all RPV steels.

An RPV is divided into different regions, each corresponding to a unique heat of steel, be it of a weld, a plate, or a forging. Figure D-1 illustrates these various regions. In the FAVOR analysis, each region is further divided into subregions of approximately constant fluence (based on the fluence maps provided by Brookhaven National Laboratory), with each subregion having an approximately constant value of fluence within it. The average amounts of copper, nickel, and phosphorus for each of these regions is based on the information in the RVID database, and is summarized in Appendix C. This appendix employs available data to estimate the distribution of chemical composition about these best-estimate values.

To appropriately model the uncertainty in chemical composition, composition variability is defined at two different levels:

(1)   The possible composition variability within a region is defined based on multiple measurements taken from various locations within a heat of steel (see Section D.1).

(2)   Regarding variability within a subregion, in any given RPV, FAVOR simulates the existence of thousands of flaws. It is therefore possible that two (or more) of these flaws will be simulated to exist within the same subregion. The greatest physical separation that these multiple flaws could have from each other is on the order of 3 in., because (1) once a flaw is placed within a subregion, its location is specified only by its location through the wall thickness, (2) flaws are simulated to exist only in the inner 3/8-T of the vessel wall, and (3) pressurized-water reactor (PWR) vessel walls tend to be about 8 in. thick. Thus, for subregions, the possible composition variability is defined based on multiple measurements made close together, as detailed in Section D.2.

D.1 Variability Within a Region

D.1.1 Welds

D.1.1.1 Copper and Nickel

The raw data used to quantify the variability of copper and nickel within a particular weld region was obtained from reports published by the Combustion Engineering Owners Group (CEOG) and the Babcock & Wilcox Nuclear Owners Group (B&WOG) that organize individual measurements of chemical composition in terms of the hierarchy illustrated in Figure D-2 (see CEOG). Within each heat of material, data may be available for several different weld-pieces. A weld-piece is a separately identifiable weld, such as a nozzle dropout, a surveillance weld, and a weld-qualification block. For each weld-piece, some number of independent measurements of chemical composition is made. This appendix reports mean and standard deviation values at the heat level. These parameters are defined according to the following four steps:

(1)   Identify all of the independent measurements and weld-pieces associated with a particular weld-wire heat.

(2)   Determine the mean copper, nickel, and phosphorus value for each weld-piece as the average of all of the independent measurements for that weld-piece.

(3)   Determine the mean copper, nickel, and phosphorus value for the heat as the average of all of the weld-piece means (calculated in step 2).

(4)   Determine the standard deviation of copper, nickel, and phosphorus values for the heat as the standard deviation of all of the weld-piece means (calculated in step 2).

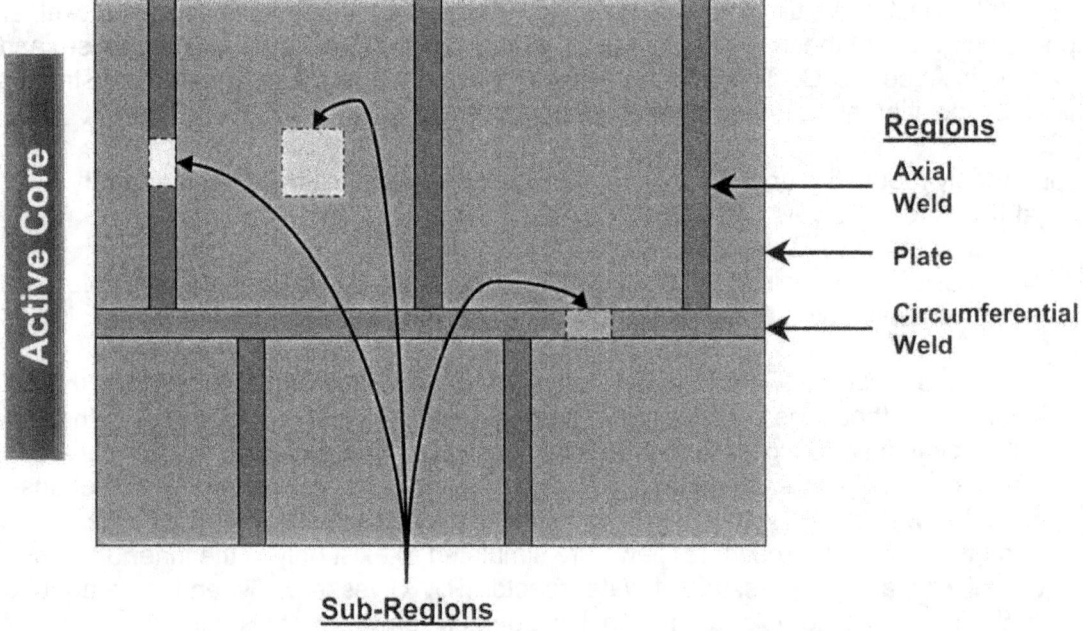

**Figure D-1 Designation of material regions and subregions
in an unwrapped view of an RPV**

This procedure weights the data from each weld-piece equally regardless of the number of independent measurements made on that weld-piece. Table D-1 and Table D-2 provide the data for copper and nickel, respectively. Statistical representations of these data are provided in Figure D-3 and Figure D-4. These fits are as follows:

•   For copper, the best estimate on the standard deviation (σ_{Cu}) is 0.167 of the mean copper value taken from Appendix C (i.e., $\sigma_{Cu} = 0.167 \cdot \mu_{Cu}$). The distribution of σ_{Cu} about this best estimate is a normal distribution. Values of the standard deviation on σ_{Cu} are as follows:

$$\sigma_{\sigma_{Cu}} = MIN\left\{\frac{0.167}{3.09023} \cdot \mu_{Cu}, 0.0185\right\}$$

This distribution is illustrated in Figure D-3. In FAVOR the standard deviation on copper for a particular heat should be simulated by drawing randomly from the distribution illustrated in Figure D-3. The standard deviation depends on the heat mean copper value, as illustrated in the figure.

• For nickel, the best estimate on the standard deviation (σ_{Ni}) is 0.029, and is independent of the mean nickel value taken from Appendix C. The distribution of σ_{Ni} about this best estimate is a normal distribution. The standard deviation on σ_{Ni} is 0.0165. Figure D-7 illustrates this distribution, truncated at the 5th and 95th percent quantiles[§§§]. In FAVOR, the standard deviation on nickel for a particular heat should be simulated by drawing randomly from the distribution illustrated in Figure D-3. This standard deviation is independent of the heat mean nickel value, as illustrated in the figure.

It may also be pointed out that the uncertainties on Cu and Ni represented in Figure D-3 and Figure D-4 (respectively) agree with the summary information on chemical composition uncertainty reported in a comprehensive survey of chemical variability in A533B plate published in the mid 1970s [Kawasaki 74, Kunitake 75].

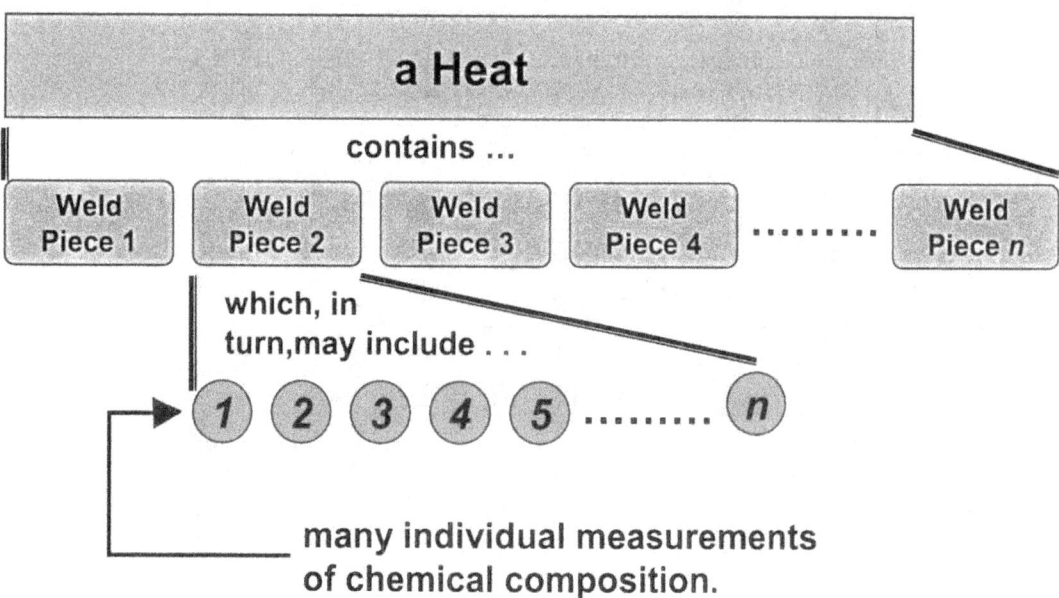

Figure D-2 Hierarchy for composition measurements

[§§§] Here 5/95 percent truncation limits are selected rather than the 1/99 percent values used in the remainder of the document to avoid simulation of negative values of standard deviation.

Table D-1 Copper Data

Vendor	PTS Plant?	Heat #	Number of Weld-Pieces	For This Heat	
				Mean	Std. Dev.
CE	Y	33A277	25	0.258	0.048
CE	Y	90136	15	0.269	0.076
BW	Y	61782	13	0.232	0.042
BW	N	72105	12	0.323	0.048
BW	Y	71249	10	0.234	0.046
CE	Y	W5214	10	0.225	0.062
CE	N	51912	10	0.156	0.012
CE	N	2P5755	10	0.210	0.036
CE	N	90099	9	0.209	0.043
CE	N	5P5622	9	0.153	0.031
BW	Y	72445	8	0.218	0.029
BW	Y	299L44	8	0.336	0.062
CE	N	4P6519	8	0.133	0.049
CE	N	1P3571	7	0.295	0.078
BW	N	406L44	7	0.270	0.033
BW	Y	8T1762	6	0.192	0.023
CE	Y	27204	6	0.203	0.020
CE	N	10137	6	0.216	0.010
CE	N	51874	6	0.147	0.034
CE	N	1248	6	0.206	0.035
BW	N	821T44	6	0.237	0.033
CE	Y	21935	5	0.183	0.033
BW	N	72442	5	0.260	0.033
CE	N	86054B	5	0.214	0.023
CE	N	1P2815	5	0.316	0.093
CE	Y	305414	4	0.337	0.023
BW	N	8T1554	4	0.160	0.019
CE	N	6329637	4	0.205	0.026
CE	Y	12008,20291	3	0.199	0.037
CE	Y	34B009	3	0.192	0.011
CE	Y	305424	3	0.289	0.019
BW	N	1P0815	3	0.167	0.059
BW	N	T29744	3	0.207	0.037
CE	N	12008,21935	3	0.213	0.011
CE	N	13253	3	0.221	0.071
BW	N	1P0661	2	0.165	0.025
CE	N	20291	2	0.191	0.043
CE	N	12008,305414	2	0.300	0.028
BW	Y	1P0962	1	0.210	0.033
BW	N	8T3914	1	0.180	
CE	N	3277	1	0.247	
CE	N	51989	1	0.170	
CE	N	12008,13253	1	0.210	

Table D-2 Nickel Data

Vendor	PTS Plant?	Heat #	Number of Weld-Pieces	For This Heat Mean	For This Heat Std. Dev.
CE		4P6052	33	0.049	0.027
CE		3P7317	30	0.067	0.031
CE		4P7869	23	0.095	0.025
BW	Y	61782	12	0.516	0.053
BW		72105	12	0.578	0.020
BW	Y	71249	10	0.590	0.033
CE		90077	10	0.055	0.017
CE		2P5755	10	0.058	0.008
BW	Y	72445	9	0.543	0.057
BW	Y	299L44	8	0.676	0.038
CE		83650	8	0.087	0.027
CE		89476	8	0.069	0.023
CE		89833	8	0.054	0.023
CE		90130	8	0.133	0.073
CE		4P6519	8	0.060	0.017
CE		83642	7	0.078	0.027
CE		83653	7	0.102	0.035
CE		88114	7	0.187	0.026
CE		90071	7	0.074	0.032
CE		1P3571	7	0.755	0.045
BW		406L44	7	0.589	0.006
CE	Y	33A277	6	0.165	0.013
BW	Y	8T1762	6	0.567	0.059
CE		10120	6	0.063	0.037
CE		90069	6	0.076	0.059
CE		90146	6	0.082	0.038
CE		90209	6	0.111	0.042
CE		5P5622	6	0.077	0.031
CE		86054B	6	0.046	0.004
BW		821T44	6	0.628	0.009
CE	Y	27204	5	1.018	0.047
CE		83637	5	0.066	0.033
CE		83640	5	0.088	0.031
CE		87005	5	0.151	0.032
CE		1P2815	5	0.724	0.021
CE		BOLA	5	0.910	0.020
BW		72442	5	0.602	0.020
CE	Y	305414	4	0.609	0.022
CE		10137	4	0.043	0.026
CE		51874	4	0.038	0.005
CE		51912	4	0.059	0.025
CE		83648	4	0.130	0.018
CE		90144	4	0.043	0.006
BW		8T1554	4	0.568	0.068
CE	Y	305424	3	0.630	0.018
CE	Y	12008,20291	3	0.846	0.026
CE		13253	3	0.732	0.007
CE		88112	3	0.188	0.045
CE		HODA	3	0.938	0.051
BW		1P0815	3	0.523	0.037

Vendor	PTS Plant?	Heat #	Number of Weld-Pieces	For This Heat	
				Mean	Std. Dev.
BW		T29744	3	0.653	0.017
CE	Y	21935	2	0.704	0.034
CE	Y	90136	2	0.070	0.000
CE		12420	2	1.023	0.033
CE		12008,305414	2	0.765	0.035
CE		12008/27204	2	0.980	0.000
BW		1P0661	2	0.640	0.010
BW		8T3914	2	0.625	0.035
BW	Y	1P0962	1	0.640	
CE		9565	1	0.080	
CE		20291	1	0.737	
CE		51989	1	0.165	
CE		12008,13253	1	0.083	
CE		12008,21935	1	0.867	
CE		12008/305424	1	0.810	
CE		1P2809	1	0.770	
CE		39B196	1	1.200	
CE-Ni+		1248	4	1.073	0.142
CE-Ni+		1248/661h577	2	1.105	0.021
CE-Ni+	Y	34B009	3	0.888	0.299
CE-Ni+	Y	W5214	12	1.025	0.137

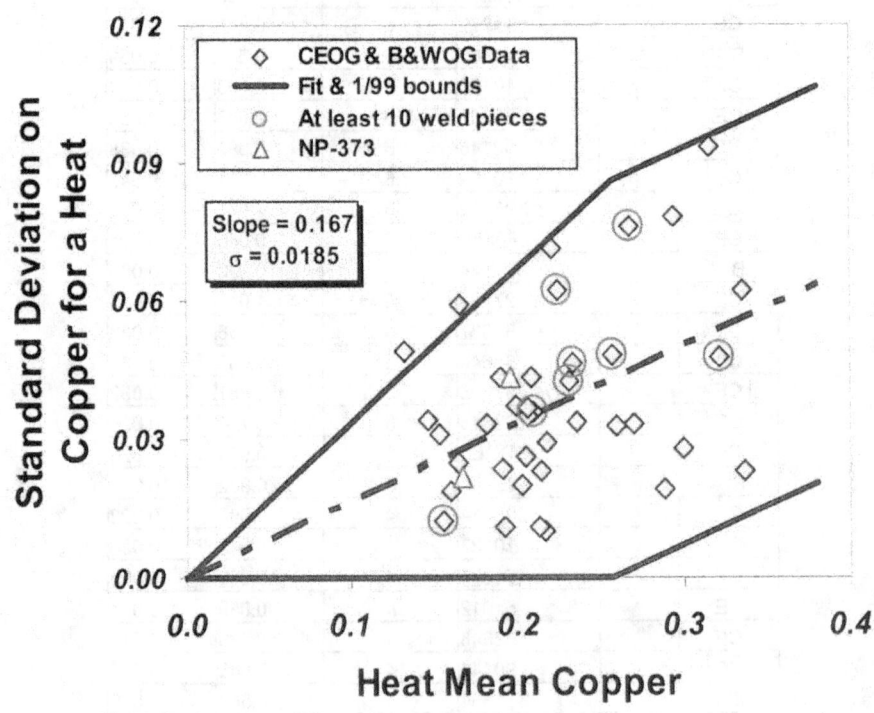

Figure D-3 Copper variability within a region for welds

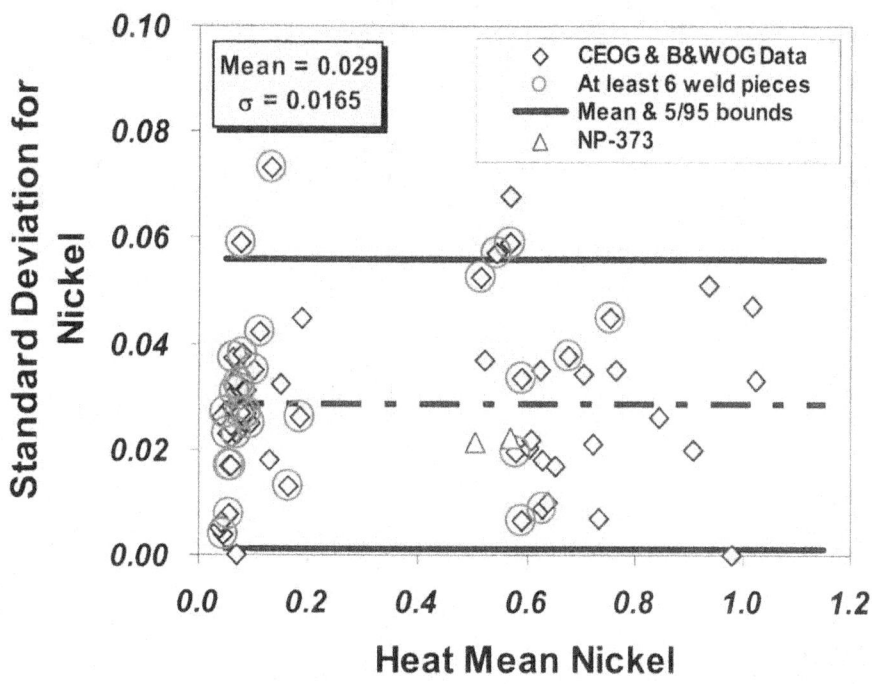

Figure D-4 Nickel variability within a region for nonnickel addition welds

D.1.1.2 Phosphorus

The data used to quantify the variability of phosphorus within a particular weld region was obtained from a 1977 Electric Power Research Institute (EPRI) report concerning a comprehensive chemical survey of a single Linde-80 weldment made by Babcock & Wilcox (see VanDerSluys 77). Figure D-5 provides the welding details and illustrates the chemical sampling plan used in this study. In total, 56 independent measurements of composition were made on the weld metal, while 35 were made in the surrounding A503 Cl. 2 forging. Figure D-3 and Figure D-4 show the copper and nickel data from these two weld-wire heats analyzed in the manner detailed in Section D1.1.1, overlaid on the larger dataset for copper and nickel. This comparison suggests that the data reported in EPRI NP-373 are similar to those available for the larger population of domestic RPV welds. Figure D-6 illustrates the phosphorus data for both the forging and for the two weld heats. The estimated standard deviation values for weld-wire heats A and B are 0.0010 and 0.0014, respectively, while the forging has an estimated standard deviation for phosphorus of 0.0016. Lacking more detailed information, it is recommended that FAVOR adopt the same standard deviation for phosphorus in all product forms, that being the average of these three experimental observations, or 0.0013. This value of 0.0013 for the standard deviation on Phosphorus agrees with the summary information on chemical composition uncertainty reported in a comprehensive survey of chemical variability in A533B plate published in the mid 1970s [Kawasaki 74, Kunitake 75].

The use of the weld and forging data together to establish a generic statistical distribution for phosphorus is justified since phosphorus is an impurity element and is not added intentionally to any product form.

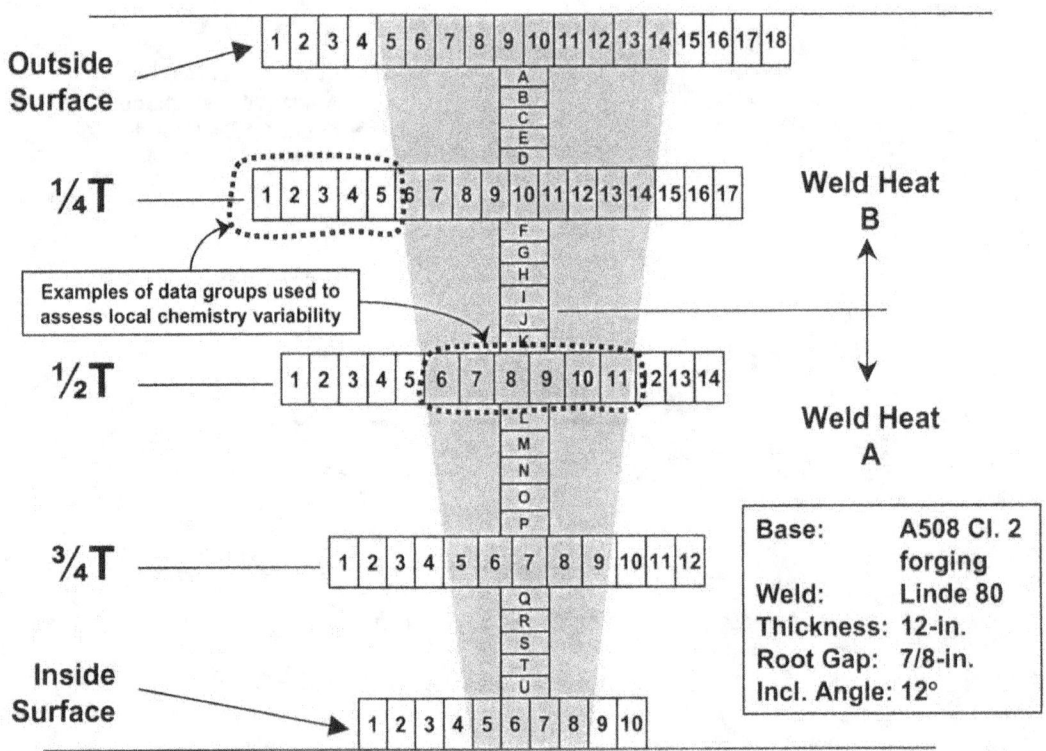

Figure D-5 Chemistry sampling plan from EPRI NP-373

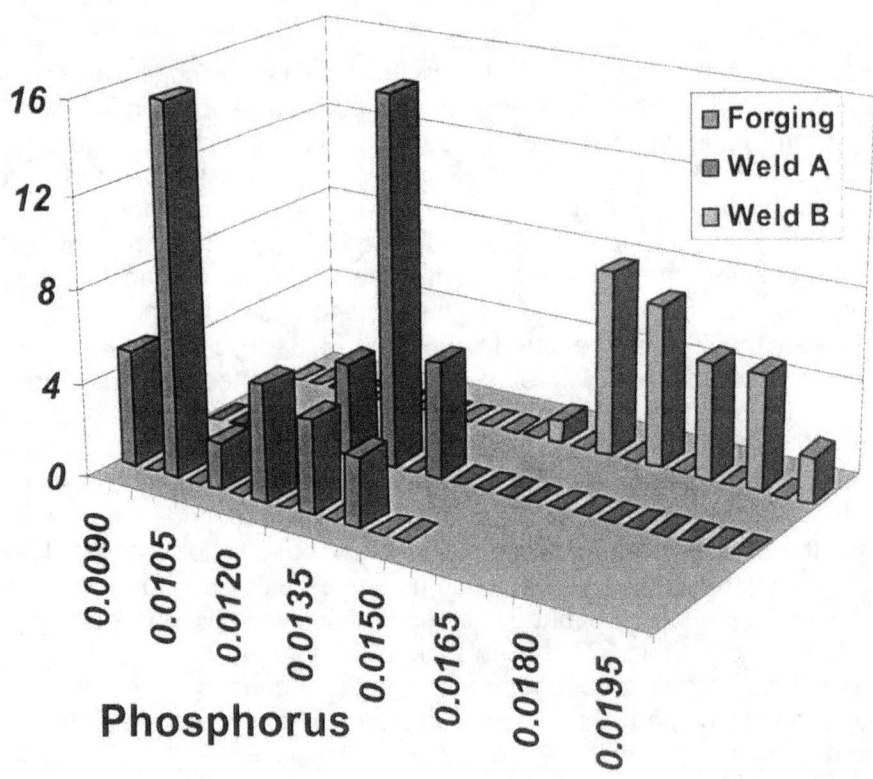

Figure D-6 Phosphorus data reported in EPRI NP-373 (the vertical axis reflects the number of independent measurements made)

D.1.2 Plates and Forgings

The data reported in EPRI NP-373 is the most detailed chemical survey of a domestic production RPV weldment that the U.S. Nuclear Regulatory Commission staff has been able to locate. For this reason, the distributions of nickel and copper determined from the 35 composition measurements made in the forging are used to assess the statistical distributions that should be assumed for copper, nickel, and phosphorus for both plates and forgings. Figure D-7 and Figure D-8 summarize the copper and nickel data, respectively (the phosphorus data were presented previously in Figure D-6). Based on these data, the following distributions are recommended for use in FAVOR to represent the chemical composition variability in all plate and forging regions:

- for copper, normal with a standard deviation of 0.0073
- for nickel, normal with a standard deviation of 0.0244
- for phosphorus, normal with a standard deviation of 0.0013 (see Section D.1.1.2)

D.2 Variability within a Subregion

To quantify the variability in copper, nickel, and phosphorus that could be expected to occur should FAVOR simulate more than one flaw to exist within the same subregion, data sets were assembled from the literature in which multiple measurements of chemistry were made close together (i.e., within the area covered by a few square inches). The following data sources were identified:

- CE-NPSD-944—Five measurements of weld chemistry (copper and nickel) were made at the 1/4-T location on eight different samples of weld, these samples having been removed from a total of seven weld wire heats.

- EPRI NP-371—As illustrated in Figure D-5, many groupings of chemistry measurements taken from this comprehensive study of chemistry can be used to assess the local variability of plate and weld chemistry.

In order to use all of these data together, the mean values of copper, nickel, and phosphorus were first calculated for each local grouping. The deviation of each weld measurement from this local mean was then calculated, and a normal distribution fit to the deviation values to quantify the local variability in chemistry. Table D-3 summarizes these standard deviations, while Figure D-9 provides histograms of the underlying data. Should FAVOR simulate multiple flaws to exist within the same subregion, normal distributions having the standard deviations from Table D-3 should be sampled, and this sampled value added to the previously simulated mean values of chemistry for that subregion.

D.3 References

CEOG	Combustion Engineering Owners Group, "Fracture Toughness Characterization of C-E RPV Materials," Draft Report, Rev. 0, CE NSPD-1118, 1998.
Kawasaki 74	Kawasaki, M. and Fujimura, T., "The Present State of Manufacturing Light Water Reactor Pressure Vessels in Japan (Part 1. Heavy Section RPV Steels)," Japan Atomic Energy Research Institute Report JAERI-M-5809, July 1974.

Kunitake 75	Kunitake, H., et al., "Manufacture and Properties of Heavy Section Steel Plates for Nuclear Reactor Pressure Vessels," Nippon Steel Corporation, Nippon Steel Technical Report Overseas No. 7, November 1975.
RPVDATA	Griesbach, T.J., and J.F. Williams, "User's Guide to RPVDATA, Reactor Vessel Materials Database," Westinghouse Energy Systems Business Unit, WCAP-14616, 1996.
RVID2	U.S. Nuclear Regulatory Commission Reactor Vessel Integrity Database, Version 2.1.1, July 6, 2000.
VanDerSluys 77	VanDerSluys, W.A., Seeley, R.R., and Schwabe, J.E., "An Investigation of Mechanical Properties and Chemistry within a Thick MnMoNi Submerged Arc Weldment," Electric Power Research Institute Report, EPRI NP-373, February 1977.

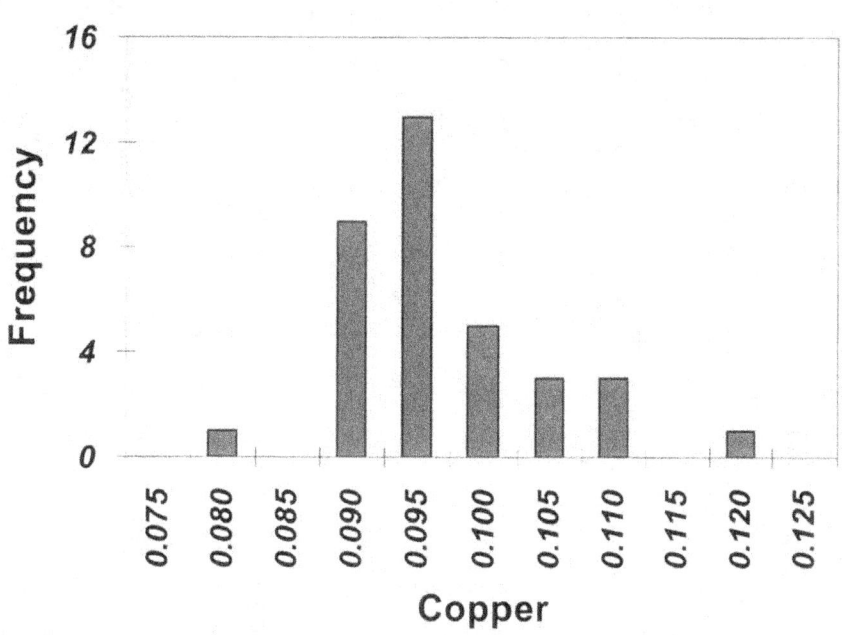

Figure D-7 Copper data reported in EPRI NP-373 for an A508 Cl. 2 forging (the vertical axis reflects the number of independent measurements made)

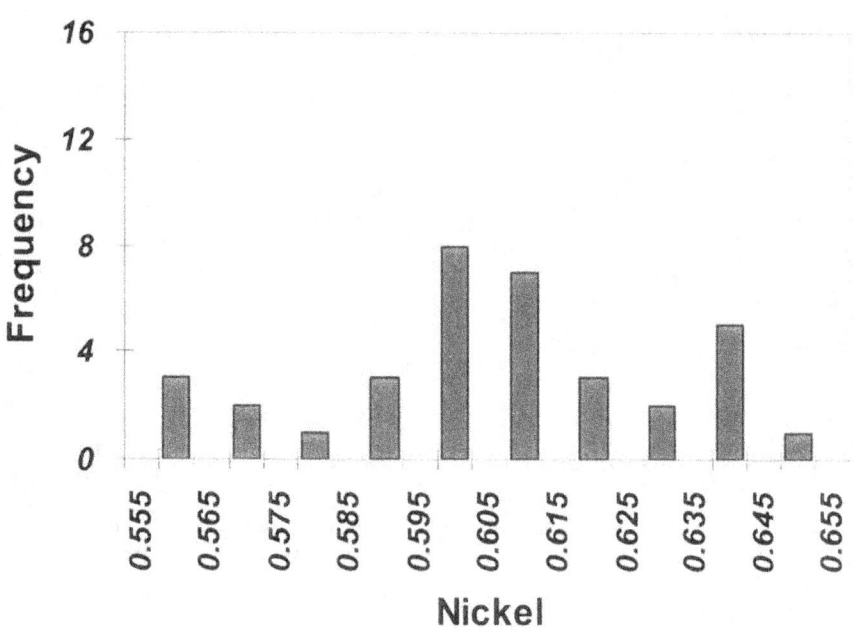

Figure D-8 Nickel data reported in EPRI NP-373 for an A508 Cl. 2 forging (the vertical axis reflects the number of independent measurements made)

Table D-3 Standard Deviations to Quantify Local Variability

	For Welds	For Plates and Forgings
Copper	0.0131	0.0035
Nickel	0.0119	0.0124
Phosphorus	0.0008	0.0005

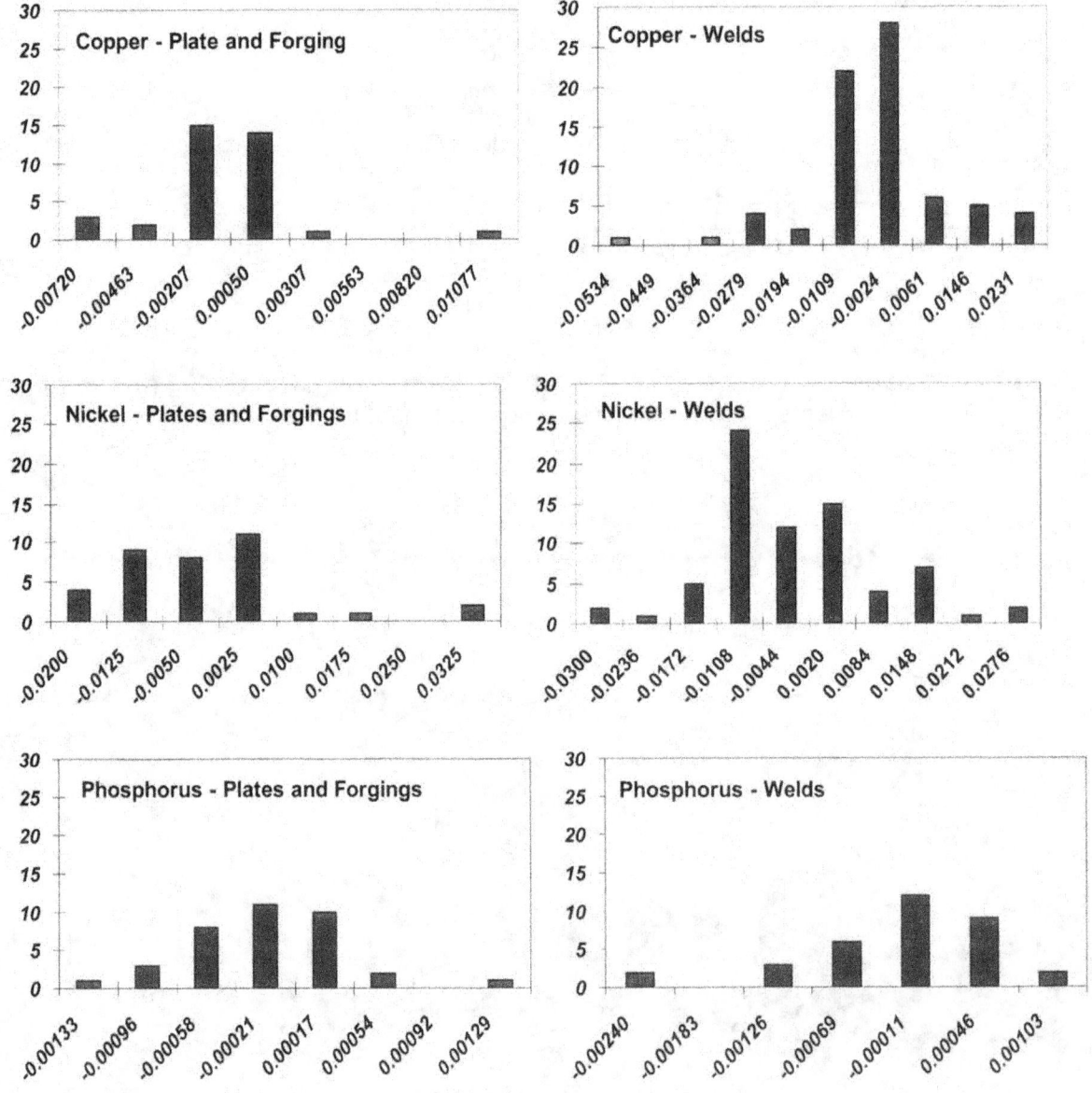

Figure D-9 Local chemistry variability histogram

APPENDIX E
DATA SOURCES FOR UPPER-SHELF TOUGHNESS CHARACTERIZATION (FROM ERICKSONKIRK 04b)

70.   Hiser, A.L., et al., "J-R Curve Characterization of Irradiated Low Upper Shelf Welds," USNRC, NUREG/CR-3506, 1984.

71.   McCabe, D.E., "Evaluation of WF-70 Weld Metal from the Midland-1 Reactor Vessel," USNRC, NUREG/CR-5736, 2000.

72.   McGowan, J.J., et al., "Characterization of Irradiated Current Practice Welds and A533 Grade B Class 1 Plate for Nuclear Pressure Vessel Service," USNRC, NUREG/CR-4880, 1988.

73.   Hawthorne, J.R., et al., "Investigations of Irradiation-Anneal-Reirradiation (IAR) Properties Trends of RPV Welds: Phase 2 Final Report," US-NRC, NUREG-CR/5492, 1990.

74.   Bryan, R.H., et al., "Pressrized Thermal Shock Test of 6-in. Thick Pressure Vessels. PTSE-2: Investigation of Low Tearing Resistance and Warm Prestressing," USNRC, NUREG/CR-4888, 1987

75.   Onizawa, K. and Suzuki, M., "Comparison of Transition Temperature Shifts Between Static Fracture Toughness and Charpy-v Impact Properties Due to Irradiation and Post-Irradiation Annealing for Japanese A533B-1 Steels," *Effects of Radiation on Materials: 20th International Symposium, ASTM STP 1405*, S. T. Rosinski, M. L. Grossbeck, T. R. Allen, and A. S. Kumar, Eds., American Society for Testing and Materials, West Conshohocken, PA, 2002.

76.   Gudas, John P., "Micromechanics of Fracture and Crack Arrest in Two High Strength Steels," DTNSRDC/SME-87-20, 1987.

77.   Czyryca, E.J., et al., "Fracture Toughness of HSLA-100, HSLA-80, and ASTM A710 Steel Plate," DTRC-SME-88/64, 1990.

78.   Kirk, M.T., "Applicability of ASTM A710 Grade A Class 3 (HSLA-80) Steel for use as Crack Arrestors," DTNSRDC/SME-87-54, 1987.

79.   Natishan, MarjorieAnn Erickson, "Mechanisms of Strength and Toughness in a Microalloyed, Percipitation Hardened Steel," DTRC/SME-88-81, 1988.

80.   M. A. Sokolov, R. K. Nanstad, I. Remec, C. A. Baldwin, and R. L. Swain, "Fracture Toughness of an Irradiated, Highly Embrittled Reactor Pressure Vessel Weld," ORNL/TM-2002/293.

NRC FORM 335
(2-89)
NRCM 1102,
3201,3202

U.S. NUCLEAR REGULATORY COMMISSION

BIBLIOGRAPHIC DATA SHEET

(See instructions on the reverse)

1. REPORT NUMBER
(Assigned by NRC, Add Vol., Supp., Rev.,
and Addendum Numbers, if anv.1

NUREG-1807

2. TITLE AND SUBTITLE

Probabilistic Fracture Mechanics—Models, Parameters, and Uncertainty Treatment Used in FAVOR Version 04.1

3. DATE REPORT PUBLISHED

MONTH	YEAR
June	2007

4. FIN OR GRANT NUMBER

5. AUTHOR(S)

M. EricksonKirk, B.R. Bass, T. Dickson, C. Pugh, T. Santos, P. Williams

6. TYPE OF REPORT

Technical

7. PERIOD COVERED *(Inclusive Dates)*

6-99 to 6-06

8. PERFORMING ORGANIZATION - NAME AND ADDRESS *(If NRC, provide Division, Office or Region, U.S. Nuclear Regulatory Commission, and mailing address; if contractor, provide name and mailing address.)*

Division of Fuel, Engineering, and Radiological Research
Office of Nuclear Regulatory Research
U. S. Nuclear Regulatory Commission
Washington, DC 2055-0001

9. SPONSORING ORGANIZATION - NAME AND ADDRESS *(If NRC, type "Same as above"; if contractor, provide NRC Division, Office or Region, U.S. Nuclear Regulatory Commission, and mailing address.)*

Division of Fuel, Engineering, and Radiological Research
Office of Nuclear Regulatory Research
U. S. Nuclear Regulatory Commission
Washington, DC 2055-0001

10. SUPPLEMENTARY NOTES

11. ABSTRACT *(200 words or less)*

During plant operation, the walls of reactor pressure vessels (RPVs) are exposed to neutron radiation, resulting in localized embrittlement of the vessel steel and weld materials in the core area. If an embrittled RPV had an existing flaw of critical size and certain severe system transients were to occur, the flaw could very rapidly propagate through the vessel, resulting in a through-wall crack and challenging the integrity of the RPV. The severe transients of concern, known as pressurized thermal shock (PTS), are characterized by a rapid cooling (i.e., thermal shock) of the internal RPV surface in combination with repressurization of the RPV. Advancements in our understanding and knowledge of materials behavior, our ability to realistically model plant systems and operational characteristics, and our ability to better evaluate PTS transients to estimate loads on vessel walls led the U.S. Nuclear Regulatory Commission (NRC) to realize that the earlier analysis, conducted in the course of developing the PTS Rule in the 1980s, contained significant conservatisms.

This report, which describes the technical basis for the probabilistic fracture mechanics model, is one of a series of 21 other documents detailing the results of the NRC study.

12. KEY WORDS/DESCRIPTORS *(List words or phrases that will assist researchers in locating the report.)*

pressurized thermal shock, probabilistic fracture mechanics, reactor pressure vessels;

pressurized water reactor

13. AVAILABILITY STATEMENT

unlimited

14. SECURITY CLASSIFICATION

(This Page)

unclassified

(This Report)

unclassified

15. NUMBER OF PAGES

16. PRICE